The Kasimovian Age of the Carboniferous Period

Bridging the Past to the Present

Origin, Evolution and Geological History of the Earth

More information about this series can be found at
https://novapublishers.com/product-category/series/origin-evolution-and-geological-history-of-the-earth/

Insights into Biological and Cultural Evolution

More information about this series can be found at
https://novapublishers.com/product-category/series/insights-into-biological-and-cultural-evolution/

Abir U. Igamberdiev

The Kasimovian Age of the Carboniferous Period

Bridging the Past to the Present

Copyright © 2026 by Nova Science Publishers, Inc.
DOI: https://doi.org/10.52305/WDZU5148

All rights reserved. No part of this book may be reproduced, stored in a retrieval system or transmitted in any form or by any means: electronic, electrostatic, magnetic, tape, mechanical photocopying, recording or otherwise without the written permission of the Publisher.

We have partnered with Copyright Clearance Center to make it easy for you to obtain permissions to reuse content from this publication. Simply navigate to this publication's page on Nova's website and locate the "Get Permission" button below the title description. This button is linked directly to the title's permission page on copyright.com. Alternatively, you can visit copyright.com and search by title, ISBN, or ISSN.

For further questions about using the service on copyright.com, please contact:
Copyright Clearance Center
Phone: +1-(978) 750-8400   Fax: +1-(978) 750-4470   E-mail: info@copyright.com.

NOTICE TO THE READER

The Publisher has taken reasonable care in the preparation of this book, but makes no expressed or implied warranty of any kind and assumes no responsibility for any errors or omissions. No liability is assumed for incidental or consequential damages in connection with or arising out of information contained in this book. The Publisher shall not be liable for any special, consequential, or exemplary damages resulting, in whole or in part, from the readers' use of, or reliance upon, this material. Any parts of this book based on government reports are so indicated and copyright is claimed for those parts to the extent applicable to compilations of such works.

Independent verification should be sought for any data, advice or recommendations contained in this book. In addition, no responsibility is assumed by the Publisher for any injury and/or damage to persons or property arising from any methods, products, instructions, ideas or otherwise contained in this publication.

The Publisher assumes no responsibility for any statements of fact or opinion expressed in the published contents.

This publication is designed to provide accurate and authoritative information with regard to the subject matter covered herein. It is sold with the clear understanding that the Publisher is not engaged in rendering legal or any other professional services. If legal or any other expert assistance is required, the services of a competent person should be sought. FROM A DECLARATION OF PARTICIPANTS JOINTLY ADOPTED BY A COMMITTEE OF THE AMERICAN BAR ASSOCIATION AND A COMMITTEE OF PUBLISHERS.

Additional color graphics may be available in the e-book version of this book.

Library of Congress Cataloging-in-Publication Data

ISBN: 979-8-90134-059-2 (Softcover)
ISBN: 979-8-90134-082-0 (eBook)

Published by Nova Science Publishers, Inc. † New York

*Dedicated to my grandparents and first teachers
Fyodor Andreevich Dashkov and Maria Alexandrovna Dashkova*

Contents

Preface		xi
Chapter 1	**The Oka River**	1
Chapter 2	**In the Depths of the East European Platform**	5
	2.1. Early History of the Earth	5
	2.2. Formation of the East European Platform	9
Chapter 3	**The Kasimovian Age of the Carboniferous Period**	13
	3.1. Flora and Fauna of the Carboniferous Period	13
	3.2. Divisions of the Carboniferous Period	16
	3.3. The Geologists – Explorers of the Upper Carboniferous Stratigraphy in Central Russia	17
	3.4. Geological History of the Area Near the Town of Kasimov	18
Chapter 4	**Terrestrial Flora at the End of the Carboniferous Period**	23
	4.1. Climate Trends in the Late Carboniferous and Early Permian Period	23
	4.2. Joggins Fossil Cliffs	26
Chapter 5	**Marine Fauna at the End of the Carboniferous Period**	29
	5.1. Brachiopods	29
	5.2. Sea Lilies	30
	5.3. Foraminifera	31
	5.4. Corals	32

Chapter 6	**Role of Land Plants in the Atmospheric O_2 Buildup and CO_2 Depletion** 33	

6.1. Gaseous Content of the Atmosphere in the Earth's History 33
6.2. Concentrations of O_2 During the Phanerozoic Eon 34
6.3. Concentrations of CO_2 During the Phanerozoic Eon and Limestone Deposition 36
6.4. Land Plant Photosynthesis and the O_2/CO_2 Balance 37
6.5. Difference in Photosynthesis of Land Plants and Algae 38
6.6. Respiration and Photorespiration 40
6.7. Land Plant Photosynthesis and Photorespiration in the Regulation of O_2 and CO_2 41
6.8. The Role of Weathering by Land Plants in CO_2 Depletion in the Atmosphere 41
6.9. The O_2/CO_2 Compensation Ratio 43
6.10. Correlation of O_2 and CO_2 in the Atmosphere with the Evolutionary Process 44
6.11. Greenhouse Bursts and Their Consequences 47
6.12. Concluding Remarks 48

Chapter 7 **Events of the Quaternary Period and the Role of Carboniferous Limestone** 51
7.1. Glacial-Interglacial Oscillations in the Quaternary Period 51
7.2. Quaternary Development of the Oka Floodplain 54

Chapter 8 **The Oka Flora on the Carboniferous Limestone** 59
8.1. The Explorers of the Oka Flora 59
8.2. The Postglacial Spreading of Plants 65

Chapter 9 **Holocene: First Human Settlements and Neolithic Agriculture** 69

Chapter 10 **Anthropocene and Recent Carboniferous Limestone Developments** 75

Chapter 11 **The Town of Kasimov** 79

Afterword	85
References	87
Appendix I: Fyodor Dashkov: History of the Village of Pervo	97
Appendix II: From My Childhood Memories	103
Appendix III: Issues of the Oka Vegetation Distribution: The Oka Flora in the Kasimov Region of the Ryazan Oblast	109
Main Findings	115
References	115
Index	117

Preface

This book brings forth the project to reconstruct the natural history of the Earth and the history of the development of human civilization on the example of the area of the banks of the Oka River, where the author spent his childhood and where he developed an interest in the study of natural science. You can travel in space to learn about the world and travel in time, being at one point in space. Without leaving Königsberg, Immanuel Kant created one of the greatest philosophical systems. My task is much more modest: in this book, I am tracing the history of the Earth, starting from its origin, from a single point of reference. Without expanding space, we can better understand the flow of time. The objective astronomical and geological time represents a synthesis of the individual times of the organisms that inhabited the planet and collectively created its unique history. With the advent of man, this temporal flow is intensified, the dynamic processes are increased, and a new, unique world of the noosphere is formed by integrating what has been accumulated over hundreds of millions of years of biosphere evolution, in which human consciousness became the engine of global processes. Developing the optimal and non-destructive trajectories of this movement is the primary task of civilization, and our knowledge of the natural history of the native land is the first important step toward its implementation.

Chapter 1

The Oka River

I spent my childhood on the banks of the Oka River. It turned out to be a river that allowed me to travel through the history of the Earth, discovering it in different projections: the geological history of the Earth, the history of vegetation, and the history of human civilization since the Neolithic time. Therefore, I will start my "brief history of time" with the basic information about this river. The Oka originates near the small town of Maloarkhangelsk in the Oryol region, passes through Kaluga, Tula, Moscow, Ryazan, Vladimir, and Nizhny Novgorod regions, and flows into the Volga River in Nizhny Novgorod. The length of the Oka is 1478 km, and the area of its basin is 245 thousand square kilometers. The Ryazan region is located in the middle reaches of the Oka. Here, the Oka calmly flows along the edge of the Meshcherskaya lowland and makes many kilometers of loops; among them, the Kochemary bows thirty kilometers above the town of Kasimov, with immense flood meadows. Then, closer to Kasimov at the mouth of the Gus River, the Oka turns sharply to the east and breaks through the Kasimov (Oka-Tsninsky) uplift (Makarova et al., 1999).

At Kasimov, the Oka forms a smooth bend facing north and becomes especially picturesque. The historical center of Kasimov is located at the top of this bend. After passing the Kasimov Uplift, the Oka goes to the Moksha Lowland, and after merging with the Moksha, it turns north and flows towards the Volga. The city of Kasimov lies on the high left bank of the Oka. Usually, according to Karl Baer's law (Goudie, 2004), the right bank is steep, but in the area of Kasimov, Oka breaks through the Kasimov coal ridge, which forms the basis of the Oka-Tsninsky uplift. Here, the Oka loses its wide floodplain and flows in a narrow passage lined with limestone rocks, with high banks replete with outcrops of white stone (limestone). Below Kasimov, the right bank again becomes high, and the left bank is occupied by flood meadows and further (2-3 km from the coast) – a sandy terrace with a pine forest (Figure 1). This is the terrain near the village of Pervo, 15 kilometers below Kasimov along the Oka (counting from the pier, which is no longer there, but 12 kilometers, if you count from the lower part of the city, Stary Posad, and through the meadows in a straight line – only 8 kilometers). This is where I spent my childhood: a steep right bank, on which the village of Pervo is located, opposite flood meadows, and an old riverbed connecting with the Oka one and a half kilometers below, opposite the village of Vasilyovo.

Below, two kilometers away, after the village of Maltsevo, the Oka branches into two channels, between which there is a picturesque Maltsevsky Island, two kilometers long. Further, the right bank is covered with forest, and to the south of the village of Balushevo-Pochinki, the left bank (the village of Shcherbatovka) becomes steep again. Above the village of Pervo (three kilometers away), the Tashenka River flows into the Oka. The height of the Oka riverbed in this place is 82 meters above sea level, and the steep bank rises to 167 meters above sea level (near the village of Savino). In this place during my childhood, there was an observation tower.

Figure 1. View from the Baishevsky Neolithic settlement to the Oka, flood meadows, and pine forest on the sandy terrace in the distance. Author's photo (2001).

By penetrating the Oka-Tsninsky Uplift, the Oka reveals the distant history of the Earth, returning us to the Carboniferous geological period of more than three hundred million years ago. We will consider this period later, but for now, we will return to the recent time (by geological standards), the 18th century, described by Mikhail Zagoskin. Mikhail Nikolaevich Zagoskin (1789-1852) was the founder of the Russian historical novel and a very popular writer of the first half of the 19th century. The story by Mikhail Zagoskin, "Kuzma Roshchin" (Zagoskin, 1902), takes place near the village of Pervo. Here is what he writes about the Oka and the village located eight versts[1] below Kasimov. The story begins with a description of this place.

> "People say that there is nothing more beautiful and diverse than the picturesque banks of the Svir River... I do not think, however, that its shores are more picturesque than the mountainous side of the wide Oka, especially in the area of thirty or forty versts in the vicinity of one county town. Its name, no doubt, readers will guess, if I say that once, with its vast district, it constituted a separate state and was the capital town of the tsar, but not the Orthodox. About eighty years ago, eight versts from this former throne city, on a high mountain, which almost vertically descends to the river itself, there was a vast manor house not at all beautiful in appearance... Half a verst from the manor's yard, also along the mountainous bank of the river, a long order of peasant huts ended in a large meadow, in the middle of which stood a village church sheathed with ashlar. Behind it, white huts for the priest and all the clergy were built. The village cemetery was located at a distance from the church, down the steep slope of the mountain, but two or three tombstones and a few wooden cabbage rolls towered in the churchyard itself."

Further, in the following pages, we find a description of these places at the time before Easter:

> "On the last day of Holy Week, at seven in the evening, Vladimir sat on the porch in front of his father's house. From this place, the view of all the surroundings was beautiful, but Vladimir did not look at the majestic course of the wide Oka, not at its steep banks, strewn with villages. His eyes did not stop at a remote and picturesque group of city houses, from among which rose the high tower of the Tatar mosque, which still exists."

Here is the description of the spring flood of the river:

> "It was already close to midnight. The broad Oka flowed softly; the meadow side of it, covered with a flood, presented the appearance of an immense lake, in the middle of which the bushes and trees half flooded with water were blackened in places."

---

[1] Approximately 8.5 km. One verst is the old Russian unit of length equal to 1.0668 km.

My first impressions of childhood are connected with these places described by Mikhail Zagoskin. The grandparents' house is the last one in the valley; then, there is a ravine, and behind the ravine is a village cemetery. On the other side, two hundred meters away, there is an old, beautiful, but crumbling church of the Nativity of the Virgin of the 17th century, a monument of pre-Petrine times, when the country was just preparing to become an empire. A little further is the church of Diomedes of the early 20th century, large, massive, devoid of the special beauty of forms, and built before the end of the empire. Close to, we see a completely ridiculous, non-functioning windmill with a wind vane that fell off during a storm. It represented a relic of communist construction, so remarkably described by Andrei Platonov in "The Pit," "Chevengur," and "The Juvenile Sea." I found myself in Platonov's native Voronezh region a few years later, and then visited the Potudan River described by him.

The change of styles is observed not only in the development of art but also in biological evolution, as Alexander Lyubishchev wrote, noting the change in styles from archaic to classical and then modernist in the evolution of chrysomelid beetles (Lyubishchev, 1982). From the Carboniferous "juvenile sea," vast limestone deposits with fossilized organisms remain. From the current information sea of civilization, the numerous texts will survive through which "the current of time's river"[2] (Derzhavin) will push through, and people will learn to preserve them for millions of years, fixing them on a material stronger than any stone. However, who will read them in the future, maybe a termite biting into the bust of Tiberius, as in the poem by Joseph Brodsky[3]? The sea itself does not remain juvenile; it is not preserved, and the water goes away, but the life that once flourished in it becomes fossilized, and history is recreated from the fossilized remains. Living organisms, as the accretions of biological time, are the spatial forms by which we measure geological time.

---

[2] https://ruverses.com/gavrila-derzhavin/the-current-of-time-s-river/2600/.
[3] https://www.nybooks.com/articles/1987/06/25/the-bust-of-tiberius/.

Chapter 2

In the Depths of the East European Platform

2.1. Early History of the Earth

Going deep into the history of our planet at the point where the author spent his childhood, we can observe what happened hundreds of millions of years ago, looking at samples of limestone ubiquitous on the banks of the Oka. We will return to the Carboniferous Period, to its Kasimovian Age – the geological division that lasted 3.3 million years (from 307.0 to 303.7 million years ago), and was a little more than three hundred million years away from us. It was the first of the two ages of the Late Pennsylvanian epoch of the Carboniferous period, preceded by the Moscovian and then Bashkirian Ages of the Middle Pennsylvanian and followed by the Gzhelian Age, which completes the Carboniferous Period (Menning et al., 2006). The North American subdivision of the Pennsylvanian subperiod is different (five stages), but the international timescale of the International Stratigraphic Committee follows the Russian subdivision into four stages. In the area where the town of Kasimov is now located, the organisms lived in a shallow sea and were preserved in limestone in a petrified (fossilized) state. First, we will travel in time further into the depths of geological epochs since most geological history unfolds on the East European platform (also called the Russian platform or Phenosarmatia). The latter is one of the largest relatively stable sections of the Earth's crust, finally shaped at the beginning of the Proterozoic eon, i.e., more than two billion years ago, when it started to form during the Archaean eon. Thus, the East European Platform already acquired its initial shape when the Earth's geological history had passed much less than halfway to our time.

The largest periods in geological chronology are called eons. Eons are divided into eras, eras into periods, periods into epochs, and epochs into ages (corresponding to tiers of stratigraphic deposits). The Earth was formed 4567 million years ago and received its current form after the formation of the Moon approximately 4425 million years ago, probably as a result of a collision with the protoplanet Theia, which was the size of Mars. Initially, the orbit of the Moon was located closer to the Earth, and the day may have lasted only two to six hours (Gordon, 2024). At the end of the Archean, the day was twelve hours, and in the middle of the Proterozoic, it was nearly twenty, which corresponded to about 450 days in a year. Currently, the lengthening of the day occurs at a rate of 1.7 milliseconds per century (McCarthy and Seidelmann, 2009).

The Moon has become a heavy satellite in low orbit around the Earth, which causes tides and thereby mixes the contents of the ocean over hundreds of millions of years, which has played an important role in ensuring a stable climate for Earth. It may also have played a role in the formation of lithospheric tectonics (plate movement), which ensures the reproduction of the carbon cycle on planet Earth and represents an atypical process in the universe.

In total, four eons are distinguished in the history of the Earth (Hadean, Archaean, Proterozoic, and Phanerozoic), and the first three are combined into the supereon of the Precambrian. The first eon – Hadean lasted a bit more than half a billion years. The Earth was then a moderately hot cosmic body (no higher than 200-300°C) with a homogeneous composition (no core or crust). Liquid water probably appeared around 4.4 billion years ago, when the atmosphere was denser, allowing water to remain liquid at a temperature of about 200 degrees. Later, the action of plate tectonics and oceans absorbed a significant part of the greenhouse gases of the atmosphere, leading to cooler surface temperatures and the formation of hard rocks. The relief of the Hadean resembled the meteorite-dotted surface of the Moon but was smoothed out due to continuous tidal earthquakes. It was composed of a monotonously dark gray substance, covered with a thick layer of regolith on top (Harrison, 2009).

The Archean eon spans 1.5 billion years, beginning 4000 and ending 2500 million years ago. During the Archean, life appeared on Earth. There was no photosynthesis at the beginning of the origin of life, but it arose first as bacterial photosynthesis, in which, in many cases, hydrogen sulfide is used, and sulfur is released, so it had nothing to do with the release of oxygen. Oxygen-emitting cyanobacteria became widespread closer to the end of the Archaean eon. The first anaerobic organisms of the Archaean supported life through various redox reactions. The widespread methane-forming Archaea (formerly called archaebacteria) used carbon dioxide and emitted methane. They maintained a high methane content in the atmosphere, which was hundreds of times higher than the current concentration, and provided a greenhouse effect, although the Sun was dimmer than now (Kasting and Ono, 2006).

Numerous sulfur, graphite, iron, and nickel deposits were formed during the Archean. The atmosphere in the early Archean was a mixed gas-vapor mass consisting of water vapor and acidic smoke, which enveloped the entire planet in a powerful and dense layer. Then, there was a separation of the hydrosphere, which formed shallow seas filled with highly saline water. The permeability of the atmosphere to sunlight was quite weak; darkness reigned on the surface, and the high chemical activity of the Archaean atmosphere led to an active effect on the basalt surface of the Earth. Large blocks of the Earth's crust (cratons) originated in the Archaean, giving rise to the continents, microcontinents, and volcanoes. At the end of the Archean, all cratons formed a single supercontinent, Vaalbara (Zegers et al., 1998). As a result of the movement of tectonic plates, Vaalbara split, and in the Proterozoic Eon, a single supercontinent gathered several times.

The Proterozoic Eon lasted almost two billion years, from 2.5 billion to 538.8 million years ago. It ended with the beginning of the Cambrian period of the Paleozoic era of the Phanerozoic Eon, which correlated with the so-called Cambrian explosion – the appearance of a large number of multicellular organisms with solid skeletons. The main characteristic of the Proterozoic is the accumulation of oxygen, probably from small fractions of a percent at the beginning of the Proterozoic to 10-15 percent at the end. As early as the beginning of the Proterozoic, oxygen caused the so-called "Oxygen Catastrophe" (Great Oxidation Event). Still, in general, it accumulated slowly, perhaps with some oscillations, and, except for prolonged glaciations, little happened on Earth, so the long duration of its accumulation in the Proterozoic is called the "Boring Billion." It is not known exactly how much atmospheric oxygen was in the period preceding the Great Oxidation Event, but it may have been a surge of up to several percent, and then a decrease in the process of glaciation, when the intensity of photosynthesis of cyanobacteria was limited due to a thick layer of ice (Shields-Zhou and Och, 2011).

At the beginning of the Proterozoic Eon, there were no multicellular organisms. There were no eukaryotic cells with a formed nucleus and organelles. Still, there were only prokaryotes, that is, Archaea, Eubacteria, and Cyanobacteria (formerly called blue-green algae, now included with the Eubacteria), which became widespread by the beginning of the Proterozoic eon. Over hundreds of millions of years of photosynthesis, cyanobacteria released vast amounts of oxygen, which first oxidized rocks and then began to accumulate in the atmosphere. Due to this accumulation, methane, previously contained in the atmosphere in significant quantities (probably up to 1-2%), was intensively oxidized by oxygen to form carbon dioxide. This methane, supplied by volcanic eruptions since the beginning of the Earth, and then continuously produced by Archaea that existed in the Archaean eon, provided a warm climate for the Earth, even though the Sun at that time shone dimmer by 25-30%. The luminosity of the Sun increased by 6 percent every billion years. Methane is a powerful greenhouse gas providing a greenhouse effect tens of times stronger than carbon dioxide.

A global change in the composition of the Earth's atmosphere at the very beginning of the Proterozoic (Great Oxidation Event) occurred in the first period of the Paleoproterozoic era (the Siderian) more than 2.4 billion years ago. As a result, the atmosphere transformed from reducing to oxidizing, which can be traced to a sharp change in the nature of the accumulation of sedimentary rocks. Almost all metamorphic and sedimentary rocks that make up most of the Earth's crust became oxidized (Holland, 2006). The Great Oxidation Event was followed by the Huronian glaciation (named after Lake Huron, one of the Great Lakes of North America). It lasted approximately 300 million years, from 2.4 to 2.1 billion years ago. Major glaciations also occurred towards the end of the Proterozoic, in the Neoproterozoic era. During the Cryogenic period, which began 720 million years ago, there were two large glaciations – at the beginning of the Sturtian (which lasted 57 million years) and at the end, the Marinoan (22 million years), between which there was a warmer interval of about ten million years. And then (635 million years ago) began the last, warmer period of the Proterozoic (the Ediacaran), with only one short glaciation event of 340 thousand years (the Gaskiers glaciation) that occurred 579 million years ago and unlike the others, did not result in snowball Earth (most of Earth frozen over). In the Ediacaran period, large aquatic multicellular organisms devoid of a solid skeleton actively spread, the peak of the diversity of which falls 20-25 million years before the "Cambrian explosion" that began 538.8 million years ago. They could have some similarities with modern coelenterates (jellyfish, corals), possibly in symbiosis with sulfur bacteria (McIlroy et al., 2021), although their systematic position remains a matter of debate. Before their appearance, sponges already lived in the Proterozoic oceans, and multicellular algae and fungi appeared, including those living on land in water films between mineral particles in zones of partial flooding near water bodies, which led to soil formation.

It must be said that the Proterozoic glaciations were incomparably more powerful than those that occurred later in the Phanerozoic Eon, and, certainly, they are incomparable with the recent Quaternary glaciations. The Earth was covered with a layer of kilometer-thick ice and resembled Jupiter's moon Europa, which has a deep (60-150 km) ocean under a layer of ice 15-25 kilometers thick. Perhaps such ice planets are common: this is also found on Saturn's moon Enceladus, on which geysers erupt, as well as the largest moons of Jupiter, Ganymede and Callisto. There may be an ocean below the surface and the most unique satellite in the solar system – Saturn's moon Titan, which has a dense nitrogen atmosphere containing methane (5% in the near-surface layer), as well as ethane, cyanogen, and other organic compounds. During the Proterozoic glaciations, the Earth was an iceball (Snowball Earth), and only volcanic activity could lead to the appearance of ice-free areas. Otherwise, evolution in the direction of the emergence of eukaryotic cells and then multicellular organisms would hardly have been possible.

During the Proterozoic Eon, there was an intense movement of tectonic plates. As a result, the supercontinent Rodinia was formed, which existed 1100 to 720 million years ago and was surrounded by the ocean of Mirovia. Before Rodinia, there were other supercontinents: Kenorland (2.75 billion years ago) and Nuna (1.8 billion years ago). After Rodinia (before the Cambrian period), there was the supercontinent Pannotia (600-540 million years ago), and then Pangea (formed about 300 million years ago and disintegrated 175 million years ago, first into Laurasia and Gondwana and then into modern continents). Intensive mountain building (accompanied by intense erosion and a decrease in temperature due to carbon dioxide sequestration during erosion) accompanies the formation of supercontinents, causing glaciations.

The Early Proterozoic was a time of intensive accumulation of iron ore, which was also associated with oxidation processes (ferrous iron was oxidized into the insoluble trivalent ferric iron), which in the East European platform formed deposits of Kryvyi Rih and the Kursk magnetic anomaly. The Late Proterozoic includes iron ores of the Urals, as well as copper, uranium, cobalt, and tin ores of other places on the planet. The oxidation of iron, copper, zinc, and other metal ions played an important role in evolution. In particular, it led to the emergence of enzymatic mechanisms triggering the formation of a solid animal skeleton, which led to the "Cambrian explosion" (Williams, 2011). The latter may be triggered by the nearly collapse of the Earth's magnetic field in the late Ediacaran, which corresponded to a key step in the evolution of multicellularity (Tarduno, 2025).

2.2. Formation of the East European Platform

We now return to the East European Platform, representing a plain dissected by the valleys of large rivers, such as the Oka. The total area of the platform is five and a half million square kilometers. The East European Platform includes the Baltic, Ukrainian shields, and the Russian Plate. The sedimentary cover of the Russian Plate includes both marine and continental sediments. In the layer of platform sediments, two structural floors are distinguished: the lower one is a crystalline basement, and the upper one is a sedimentary cover. The foundation of the platform is formed by complexly dislocated Archaean and Lower Proterozoic metamorphic rocks. The surface of the foundation is uneven and is located at a depth of half a kilometer to five kilometers. The foundation bends in the central part of the platform and is raised along its edges, forming a cup-shaped structure (the Moscow syneclise), most powerfully filled with sedimentary rocks (Grachev et al., 2006).

The platform in the Kasimov area has a thickness of the sediments of one and a half kilometers. This includes the sedimentary rocks of the Devonian period (419-359 million years ago) of about a kilometer, not coming to the surface, sedimentary rocks of the Carboniferous period (359-299 million years ago) of 200-300 meters, through which the Oka breaks, and in some places, the deposits of the Jurassic, Cretaceous, and Neogene are preserved. The Quaternary deposits are several tens of meters (or less) on top of the Paleozoic and Mesozoic rocks. The platform includes the layer of Late Proterozoic deposits below and is founded on the crystalline basement formed in the Archaean and Early Proterozoic. The geological structure near Kasimov is characterized by the presence of limestones of the Pennsylvanian subperiod of the Carboniferous period. It will be described in more detail in the next section.

The platform got its present shape at the end of the Paleoproterozoic Era when the cratons of Fennoscandia, Sarmatia, and Volga-Uralia merged and became its foundation. Its location was then almost at the South Pole. By the end of the Cryogenic Period of the Neoproterozoic Era (approximately 750 million years ago), deep faults associated with the glaciations appeared in the platform's foundation. At that time, the emerging East European Platform was dominated by land affected by powerful weathering processes. At the end of the glaciation time, water filled the faults, in which sedimentary rocks accumulated. Coarse quartz ferruginous sands appeared from the destruction of the foundation, and the carbonaceous sediments were the waste products of planktonic microorganisms that lived in these narrow but deep reservoirs.

In the last period of the Proterozoic, the Ediacaran (635-538.8 million years ago), the East European Plate began to separate from the supercontinent Pannotia and moved slightly north from the South Pole. In the east, it was adjacent to the Siberian continent, and in the west – with Laurentia (from which North America was later formed). Then, the foundation of the platform stabilized and began to bend slowly in its central part, forming the Moscow syneclise, which was filled with sedimentary rocks. In the middle of the Ediacaran, the territory was flooded by the sea, which came from the north and was inhabited by multicellular organisms (the Ediacaran fauna), devoid of a solid skeleton (Knoll, 1991).

In the Cambrian Period (538.8-486.85 million years ago), the supercontinent Pannotia split, and the East European Plate moved from the South Pole to the mid-latitudes of the Southern Hemisphere, becoming a separate continent. Its territory was covered by a shallow sea. The sea prevailed in the next, Ordovician, period (486.85–443.1 million years ago), and in the Silurian (443.1–419.62 million years ago), when plants started to inhabit the land (although already in the Ordovician mosses began to spread on the continents); the sea has retreated and the platform was located near the equator, approaching another continent – Laurasia. In the Devonian Period (419.62–358.86 million years), these continents collided and formed Euramerica. The territory was filled with the waters of the Ural Sea (Rogers and Santosh, 2004).

Chapter 3

The Kasimovian Age of the Carboniferous Period

3.1. Flora and Fauna of the Carboniferous Period

The place where the author spent his childhood (the Kasimov district of the Ryazan region) is unique in geological, floristic, historical, and cultural terms. In the ravines leading to the Oka and on the banks of the Oka, many remains of fossil marine animals can be found. These are primarily brachiopods (invertebrates with shells well preserved in limestone), the stems of sea lilies, the skeletons of corals, and the smallest shells of the protozoans *Foraminifera*.

The time when these organisms lived is more than three hundred million years ago from ours. This was the Carboniferous Period (358.86-298.9 million years ago), in the second half of which, due to the activity of earlier land plants, the carbon dioxide content in the atmosphere decreased to 0.03-0.07% or even less, and the oxygen content reached 30-35% (Figure 2). The Earth became cold due to active photosynthesis, mountain uplift, and a weaker Sun than now, and the climate changed from the greenhouse to the glacial (approximately what has been observed in the last 1-2 million years). The carbon dioxide content has decreased tenfold over the previous 100 million years due to the active photosynthesis and root activity of terrestrial plants, which led to the replacement of rock silicates with carbonates (Igamberdiev and Lea, 2006).

On land, Carboniferous forests of tree ferns, horsetails, and lycopsids dominated. In these forests lived ancient amphibians and the first flying insects – dragonflies, which, due to the high concentration of oxygen, reached a wingspan of 90 centimeters. However, in most of the territory of the current Moscow and Ryazan regions, there were no Carboniferous forests, but there were shallow seas in which marine life was actively developing: these were the inland seas of the mainland and large ocean bays. There were islands in the seas, and rivers flowed nearby, bringing sediments from the mainland, which contributed to the rapid development of marine fauna. In some periods, when the sea receded, carboniferous forests grew, and coal was formed. There are deposits of lignite (brown coal), for example, near the city of Skopin, two hundred kilometers from Kasimov. Later (in 2021), I visited Joggins Fossil Cliffs in Nova Scotia, Canada, where the fossils of Carboniferous forests are perfectly preserved.

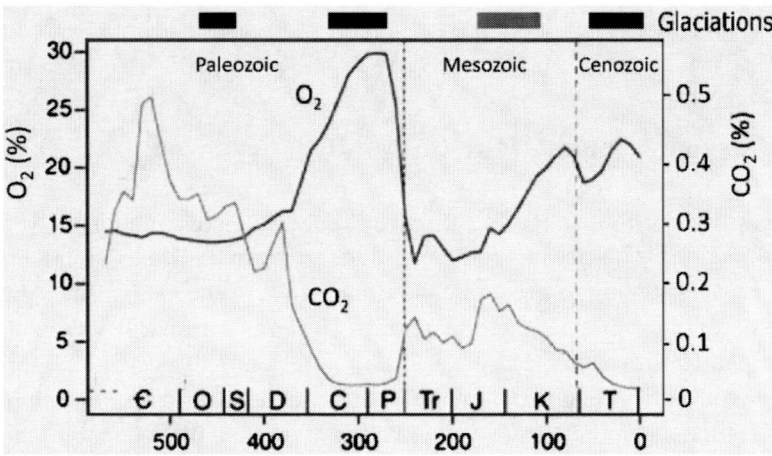

Figure 2. Fluctuations in oxygen and carbon dioxide contents in the Earth's atmosphere in the Phanerozoic. The second half of the Carboniferous period is characterized by high oxygen content (30-35%) and low carbon dioxide. Glaciation periods and eras are shown above the graph. Abbreviations of periods: €, Cambrian; O, Ordovician; S, Silurian; D, Devonian; C, Carboniferous; P, Permian; Tr, Triassic; J, Jurassic; K, Cretaceous; T, Tertiary (Paleogene and Neogene). The author's drawing.

At the bottom of the warm carboniferous sea lived large foraminifera *Fusulinida* (their calcareous shells resembled rice grains in shape and size), brachiopods (including huge giant *Productus*, whose shells were covered with long needles), a variety of echinoderms (sea lilies and sea urchins), early bivalves and gastropods, trilobites (most of them were already extinct, but one group remained).

Cephalopods with straight and curled multi-chambered shells (relatives of modern nautiluses), large sharks, coelacanths and the first bony fish – *Palaeoniscum* lived in the carboniferous sea. There were reefs built with a variety of corals and bryozoans. The climate was humid, so along the shores of this sea grew forests of giant lepidodendrons (relatives of modern spike mosses), tree-like relatives of horsetails (archaeocalamites), ferns and pteridosperms, which resembled ferns but propagated by seeds. This vegetation gave rise to coal deposits.

The Kasimovian Age of the Carboniferous Period

The protrusion of limestone in the Kasimov area has appeared because the Oka "breaks through" the East European Platform. The platform itself around Kasimov includes rocks of the crystalline basement of different ages. So, in the area of the settlement of Tuma, at a depth of about 1600 m, gneisses of the Archaean eon were discovered. The sedimentary cover of the platform is composed of sedimentary rocks of different ages: from the Upper Proterozoic to the Quaternary.

The main geological characteristics in the Kasimov region are associated with the Carboniferous period of the Paleozoic era. The Carboniferous period, by name, is associated with the accumulation of coal by plants that spread on land. However, the Kasimov area was covered by a shallow sea for a significant part of the period, and a variety of fossil organisms characterize the marine fauna that existed at that time. In the Carboniferous period, the East European Platform, as part of Euramerica, moved from the south to the zone of the northern subtropics.

Limestone deposits on the banks of the Oka near the town of Kasimov belong to the Kasimovian Age of the Pennsylvanian subperiod of the Carboniferous, corresponding to the interval of 307-303.7 million years before our time. This was a time of moderate warming after the previous cold period, which was shown by the isotopic method (reduced content of isotopes ^{18}O and ^{13}C). It was replaced by a cold snap at the end of the Kasimovian Age and in the subsequent Gzhelian Age. Later (by the second half of the Permian Period), the carbon dioxide content increased again and was quite high throughout the Mesozoic era, and only in the Cenozoic did it begin to decline again, which led to ice ages alternating approximately every 100 thousand years for the last several million years (the end of the Pliocene and the Pleistocene epochs).

By the end of the Carboniferous period, the oxygen content in the atmosphere reached 30-35%, and the carbon dioxide content fell to values close to the current ones. This was the result of photosynthesis, mainly of terrestrial plants, as well as intensive mountain building, which preceded the formation of the supercontinent Pangea, which was finally formed by the beginning of the next period, the Permian. In the process of weathering, silicates are replaced by carbonates, while atmospheric carbon dioxide is bound, and in the Carboniferous period, terrestrial plants significantly accelerated this process, releasing organic acids, settling on rocks and forming soil. All this caused a strong cooling of the climate and led to glaciations that were not as powerful as in the Proterozoic, but captured significant areas starting from the poles, and, possibly, fluctuations of glacial and interglacial periods were observed, as later in the Quaternary period.

3.2. Divisions of the Carboniferous Period

The Carboniferous period is divided into the Pennsylvanian and the Mississippian subperiods, which were previously considered as periods by American researchers. The Mississipian includes the Tournaisian, Visean, and Serpukhovian ages; the Pennsylvanian includes the Bashkirian, Moscovian, Kasimovian, and Gzhelian ages (Figure 3). The last two are united into the upper Pennsylvanian time. The Kasimovian deposits are so important and indicative that they are named after the geological age of the Carboniferous period. It is also necessary to note the important role of Russian geologists in the study of the Carboniferous period. Thus, the Kasimovian Age was separated from the Gzhelian Age by Boris Danshin (1891-1941) in 1926 and received its name from the city of Kasimov in 1949 by Georgy Teodorovich. The Moscovian Age was defined by Sergei Nikolaevich Nikitin (1851-1909) back in 1890. He also defined the Gzhelian Age in the same year (the Kasimovian Age was a part of it in his scheme) and also singled out the Serpukhovian Age as a separate one. The Bashkirian Age was defined by Sofia Semikhatova in 1934.

Period	Subperiod	Age	Time (millions of years)
Permian	Cisuralian	Asselian	younger
Carboniferous	Pennsylvanian	Gzhelian	298.9 – 303.7
		Kasimovian	303.7 – 307.0
		Moscovian	307.0 – 315.2
		Bashkirian	315.2 – 323.4
	Mississippian	Serpukhovian	323.4 – 330.3
		Viséan	330.3 – 346.7
		Tournaisian	346.7 – 358.9
Devonian	Late	Famennian	older

Figure 3. Division of the Carboniferous period into subperiods and ages (corresponding to stratigraphic layers).

3.3. The Geologists – Explorers of the Upper Carboniferous Stratigraphy in Central Russia

A few words about Russian geologists Sergei Nikitin and Boris Danshin (Figure 4). Sergei Nikolaevich Nikitin (1851-1909) made a significant contribution to the development and formation of many areas of domestic geology. His role in the formation of the state geological service of the country is significant. Boris Mitrofanovich Danshin (1891-1941) worked in the Moscow branch of the Geological Committee, participated in the creation of a geological map of the Moscow province, and then led the geological survey of European Russia, mainly of Moscow and the Moscow region. He was a member of the commission for the construction of the first stage of the Moscow Subway. In 1933, he was repressed and later released from jail. At the beginning of the war, he participated in the construction of fortifications and died in the summer of 1941.

Sergei Nikitin
1851-1909

Boris Danshin
1891-1941

Figure 4. Geologists who studied the Upper Carboniferous stratigraphy in Central Russia. The images are in the public domain.

3.4. Geological History of the Area Near the Town of Kasimov

At the beginning of the Carboniferous period (the Tournaisian Age of the Mississippian subperiod, about 350 million years ago), warm lagoons existed on the Kasimov territory. At the end of the Tournaisian Age, the territory was raised, and a continental period of development began. At the beginning of the Middle Visean Age (about 340 million years ago), the territory began to sink, and a vast coastal lowland plain was formed. Then the territory was briefly drained, and in the Serpukhovian Age (started 330.3 million years ago), the sea formed again. At the end of it (by 323.4 million years ago), the region was again drained, and the strata formed underwent weathering and erosion. A new transgression began at the end of the Bashkirian Age, that is, 315.2 million years ago. In the Moscovian Age (315.2-307.0 million years ago), the transgression of the sea continued to develop. It became deeper first, but starting from the Podolsky horizon (310 million years ago), the sea gradually became shallow, and by the end of the Gzhelian Age (298.9 million years ago) completely left this territory. This roughly corresponds to the completion of the assembly of the supercontinent Pangea. The giant ocean that surrounded Pangea and occupied half of the Earth's surface is called Panthalassa. Its large bay, called the Tethys Sea or Tethys Ocean, spread inland to what is now Eurasia. For this ocean of the Paleozoic era (Upper Carboniferous and Permian periods, 320-260 million years ago), the name Paleothetis is proposed. The territory where the town of Kasimov is now was located just on the edge of the Paleothetis Ocean, which rose and fell (perhaps in accordance with frequent glacial-interglacial fluctuations), and by the end of the Carboniferous period was isolated as the inland Central Russian Sea. It was separated from the disappearing Ural Sea by a chain of islands, which later became part of the Ural Mountains.

The limestone deposits that we will talk about belong to the period immediately preceding the formation of Pangea. In the process of the formation of Pangea from more ancient continents, mountain systems such as the Urals or the Appalachian Mountains arose at the sites of their collision. Due to the erosion that lasted for many millions of years, the Urals and Appalachian Mountains are badly damaged and low. The location of the continents in the Kasimovian Age of the Carboniferous period indicates that they almost merged into a single continent. The shallow tropical sea in the area of the present Moscow region was the bay of the Paleothetis Ocean.

In the place around Kasimov, there are limestones of the Pennsylvanian subperiod of the Carboniferous. The Kasimovian Age (the first age of the Late

Pennsylvanian) is represented around Kasimov by three horizons. Closer to the village of Pervo, there are limestones of the Middle Pennsylvanian of the Moscovian Age (the later of the four is the Myachkovsky horizon, in some places the earlier Podolsky horizon, these two belong to the Moscovian age), and in other places, there are later limestones of the Gzhelian Age (the second last stage of the Late Pennsylvanian). For orientation, we can say that the boundary between the Middle and Upper Carboniferous (between the Moscovian and Kasimovian ages) corresponds in time to 307 million years ago, and the limestones in the Kasimov area correspond to the age of 303-305 million years, the limestones of the Myachkovsky horizon – the age of 309-307 million years. This period corresponded to the extinction of many terrestrial organisms, known as the collapse of the Carboniferous rainforest, which peaked 307 million years ago, just on the border of the Moscovian and Kasimovian ages. A small extinction was also characteristic of marine animals, although the scale of the extinction was incomparable with those mass extinctions that took place at the borders of the Permian and Triassic, Triassic and Jurassic, and Cretaceous and Paleogene periods and were caused in high probability, by the fall of kilometer-sized asteroids. The extinction of organisms during the Moscovian and Kasimovian ages may be associated with climatic changes of that time, namely with cooling, accompanied by dry periods (aridification), and possible glacial-interglacial oscillations, similar to those that took place in the last two million years (in the Quaternary period).

Organogenic limestones, as well as dolomites, with inclusions and veins of gypsum along with red-colored clays, sandstones, siltstones, and marls, dominate the Kasimovian Age. The Kasimovian Age is divided (from bottom to top) into the Krevyakinsky, Khamovnichesky, and Dorogomilovsky horizons, the total thickness of which is about 70 m on the Oka in the Kasimov region, in the lower reaches of the Pet, Syntulka, and Kolpi rivers. The Podolsky and Myachkovsky horizons are also represented by limestones with interlayers of clays, marls, and dolomites, with nests of gypsum and anhydrite. They contain algae, sea lilies, foraminifera, and gastropods. Myachkovsky deposits are represented by white limestones with subordinate interlayers of dolomites and marls, with a coral-foraminiferal stratum at the base.

The limestone of the Podolsky and Myachkovsky horizons of the Moscow Age was used for the construction of the white-stone cathedrals of Vladimir and Suzdal. The Kasimov minaret of the 15th century is built of limestone excavated near the current village of Maleyevo, which is located 12 km upriver from Kasimov. At this point, the limestones of the Krevyakinsky horizon of the Kasimov tier come out, and next to it, the Myachkovsky horizon of the Moscovian Age.

Starting from the end of the Gzhelian Age (the end of the Carboniferous period, about 300 million years ago), during the Permian, Triassic, and half of the Jurassic period, the area was continental, due to powerful uplifts of the earth's crust on the East European platform at the end of the Paleozoic - the beginning of the Mesozoic. Since the end of the Gzhelian Age, the sea has receded far to the north and east, and previously deposited sediments were eroded. The limestone outcrops of the Jurassic period, north and south of Kasimov, practically do not extend to the Oka. They indicate that this area again sank into the sea in the middle part of the Jurassic period (during the Bayos, Bath, and Calloway ages, 170-163 million years ago). This corresponds to the time when the supercontinent Pangea began to break up into Laurasia and Gondwana. In this wide sea strait, connecting Tethys with the Boreal Ocean, lived bivalve and gastropod mollusks, cephalopods (ammonites and belemnites), brachiopods, sea urchins and lilies, sharks and other fish, as well as large predatory reptiles – plesiosaurs and ichthyosaurs.

In the Cretaceous period (at the beginning of the Barremian age, 125 million years ago), the area became continental, while in the Albian age of the Cretaceous period (about 110 million years ago), the sea returned and persisted during the Cenomanian age (100–94 million years ago). In the nearest places, there are no deposits of a later time up to the Neogene, when the area became land. During the Paleogene (66–23 million years ago), the continents acquired their modern shape. In the Neogene period (23–2.58 million years ago), the area became elevated and continental. At that time, the main watercourse for the territory of the southeast and center of the Russian Plain was the valley of the Paleo-Don River with a large number of tributaries. The beginning of development and the time of the foundation of the unified Volga River basin and the Paleo-Volga valley refer to the end of the Miocene and the beginning of the Pliocene (about five million years ago). At that time, the territory represented a raised, intensively dissected plain in which there was an accumulation of lacustrine-alluvial and lagoon deposits. These accumulations continued during the Quaternary period, which we will consider in the next chapters. Figure 5 presents the fossilized remains of the main representatives of the marine fauna.

The Kasimovian Age of the Carboniferous Period

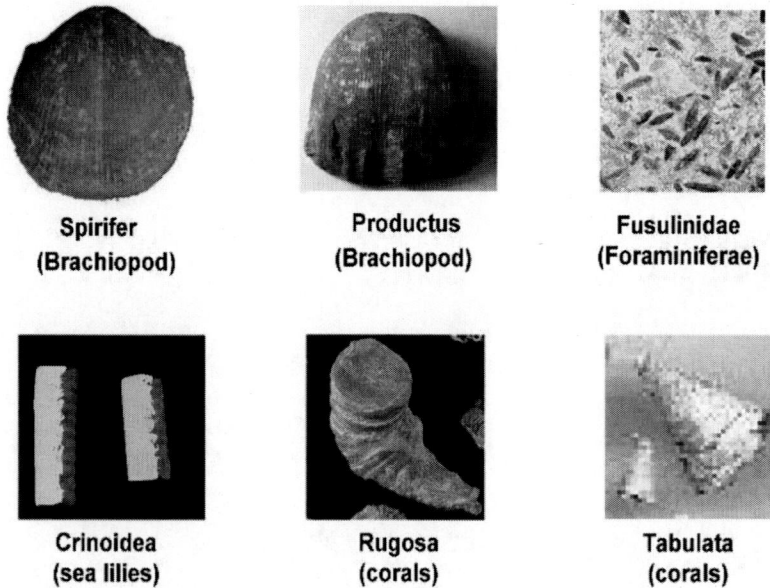

Figure 5. Main representatives of fossil marine organisms of the middle and upper Carboniferous period. All images are in the public domain.

Chapter 4

Terrestrial Flora at the End of the Carboniferous Period

4.1. Climate Trends in the Late Carboniferous and Early Permian Period

A cool climate with glaciations, high oxygen, and low CO_2 in the atmosphere characterized the time of the late Carboniferous. This icehouse climate continued during the first two ages of the subsequent Permian period, Asselian (298.9-293.5 million years ago) and Sakmarian (293.5-290.1 million years ago), which belong to the Cisuralian epoch that ended 272.3 million years ago. From 290 million years ago, CO_2 atmospheric concentration started to rise, oxygen concentration slowly decreased, and the climate turned from an icehouse to a greenhouse toward the second half of the Permian Period (Figure 6).

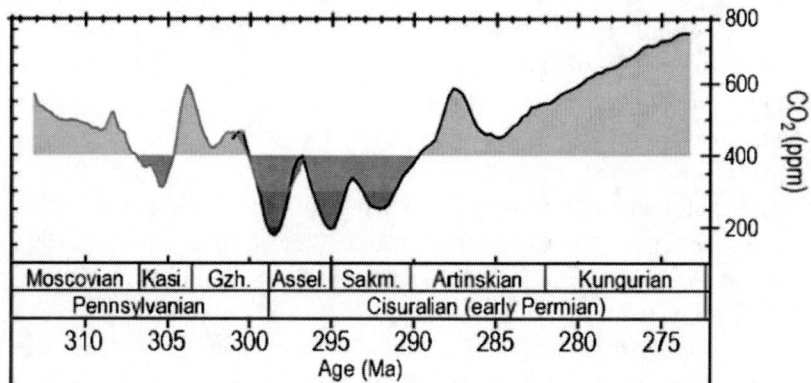

Figure 6. Late Paleozoic CO_2 concentrations according to Richey et al. (2020).

The concentrations of CO_2 were reconstructed in the paper of Richey et al. (2020). These authors presented the recalibrated data, showing that the late Paleozoic ice age was a time of dynamic glaciation and repeated ecosystem perturbation that occurred under conditions of substantial variability in atmospheric pCO_2 and O_2. Based on the numerous paleo data, they showed that the CO_2 concentration dropped during the Moscovian Age from 600 to 400 parts per million (ppm), and then further decreased to about 300 ppm in the Kasimovian Age. In the Gzhelian Age, the CO_2 concentration increased again to 400-500 ppm, followed by the drop to 200-300 ppm in most times of the Asselian and Sakmarian ages in the Permian Period. The ratio of O_2/CO_2 reached ~1000 during the times of low CO_2 in the Kasimovian Age, and then in the first two ages of Permian. At other times, it was 500 or lower. Oxygen concentration in the atmosphere reached 30-35% at that time, which was the highest in all of Earth's history. It is necessary to mention that most likely, there were shorter oscillations of the duration of tens of thousands of years, like in the recent million years; however, it is difficult to trace them to the times 300 million years ago.

The available data confirm a strong CO_2–glaciation linkage, as well as the synchronicity between major pCO_2 and O_2/CO_2 changes and compositional turnovers in terrestrial and marine ecosystems. The periods of low CO_2 corresponded to the times of glaciation.

The first glacial period occurred during the time from the Serpukhovian to the Moscovian: ice sheets expanded from a core in southern Africa and South America. A relatively warm interglacial interval named the Alykaevo Climatic Optimum spanned the end of Kasimovian and Gzhelian, preceding the later second major glacial period that occurred from the late Gzhelian across the Carboniferous-Permian boundary to the early Sakmarian of the Permian Period. During that time, ice sheets expanded from a core in Australia and India in the Southern Hemisphere.

Thus, the Kasimovian was a time of low CO_2 concentrations and a cool climate, and only by the end of this age did the CO_2 level increase to 500-600 ppm, and the climate became warmer. This lasted until the end of the Gzhelian, i.e., until the end of the Carboniferous. The climatic changes in the Kasimovian Age corresponded to the Carboniferous rainforest collapse, a minor extinction event that occurred around 305 million years ago. After this event, the composition of the forests changed from a lepidodendron-dominated forest to a forest with predominantly tree ferns and seed ferns. The previous flora temporarily survived in isolated refugia, which resulted in dwarfism, followed by the extinction of many plant and animal species. The rainforest collapse led to a decrease in oxygen concentration and a decrease in the habitat of many arthropods, including giant dragonflies. The concentration of oxygen in the seas during the abrupt warming 304 million years ago sharply decreased resulting in the anoxic conditions that significantly affected marine biodiversity (Chen et al., 2022).

Before the Carboniferous rainforest collapse, the Euramerican tropical rainforests of Euramerica were dominated by tall lycopsids (lepidodendrons), supporting rich vegetation and diverse animal life, which included giant dragonflies, millipedes, precursors of cockroaches and termites, smaller amphibians, and the first reptiles. These rainforests greatly altered the landscapes by eroding organic-rich braided river systems with many channels and stable alluvial islands. The tree-like plants led to less erosion through an increase in the complexity and diversity of root assemblages and the production of woody debris.

The Carboniferous rainforest collapse was preceded in late Moscovian by a rise of ferns followed in the early Kasimovian by a major extinction of the dominant lycopsids. Rainforests formed shrinking "islands," and by the end of the Kasimovian, they practically disappeared. Then, in the Gzhelian, they were replaced by fern tree-dominating forests growing in a drier climate. Many insects became extinct. After the collapse, the surviving rainforest "islands" developed their own unique compositions of species. Many amphibian species became extinct while the ancestors of reptiles and mammals started to diversify. The predecessors of the mammal lineage (e.g., the synapsid *Archaeothyris*) quickly recovered after the collapse. They were better adapted to the drier conditions. While amphibians breathed with both lungs and wet skin, amniotes (that included reptiles) re-evolved external plates (scales), which became more keratinized and thus conserved water. Early reptiles initially were insectivores and piscivores, but acquired new feeding strategies, including herbivory and carnivory, and evolved towards an increase in size.

The formation of coal deposits during the Carboniferous period took place simultaneously with the removal of atmospheric carbon. The Carboniferous rainforest collapse coincided with the trend of aridification. The study of paleosols (soils of past geological periods) reveals decreased hydromorphy, increased free drainage, and landscape stability. By the time of the Carboniferous rainforest collapse, the climate became cooler and drier, which led to an intense ice age with ice covering the southern part of the Gondwana supercontinent and sea levels dropping by 100 meters. At that time, rainforests survived mostly in wet valleys and then shrank further. The Carboniferous rainforest collapse affected mostly the equatorial region of Euramerica (Sahney et al., 2010). In other areas of the Earth, the Carboniferous rainforests survived until the end of the Permian, before the major and most devastating

Phanerozoic extinction event occurred 251.9 million years ago. It was preceded by the Capitanian extinction event 260 million years ago and Olson's extinction 273 million years ago, both affecting the diversity of free-sporing plants.

4.2. Joggins Fossil Cliffs

In 2021, I visited the Carboniferous site on the other side of the Earth, where the fossils of terrestrial organisms are preserved (Figure 7). This site is the Joggins Fossil Cliffs, which has a record of fossils from a rainforest ecosystem of 310-325 million years ago (Falcon-Lang, 2006), i.e., corresponding to the Bashkirian and Moscovian ages. Also, the Kasimovian age fossils can be found down to the south of the Bay of Fundy, if you walk over the 15 km Joggins coastline. Geologists explored the site starting from the 1820s, but the most important were the visits in 1842 and 1852 by Charles Lyell (1797-1875), the founder of modern geology. In his "Elements of Geology" (1871), Lyell defined Joggins fossils as "the finest example in the world." Charles Darwin, in the "Origin of Species," referred to the fossils from Joggins. They played a role in the Great Oxford Debate of 1860 on evolution between Bishop Wilberforce and Thomas Huxley. Further, the Nova Scotian geologist Sir William Dawson (1820–1899), to whom Charles Lyell was a mentor and friend, studied the Joggins Cliffs.

Figure 7. Joggins Fossil Cliffs and a fossilized calamite stem. Author's photos (2021).

Among plant organisms at the Joggins site are lepidodendrons, sigillarias, calamites, and some ferns that flourished before the Carboniferous rainforest collapse during the Bashkirian and Moscovian ages. The time of collapse could be observed in the southern part of the Joggins, which is less accessible.

Animal species include early amphibians, fish species (including coelacanths), and many arthropod species. William Dawson discovered in the Joggins the earliest known sauropsid (reptile) in the history of life, *Hylonomus lyelli*. The Joggins Fossil Cliffs represent a period when a tropical rainforest covered Nova Scotia. Slightly more recent fossils of the Kasimovian age down to the south indicate the collapse of these rainforests (Carboniferous Rainforest Collapse) that occurred 305 million years ago. After this extinction event, fern trees mostly replaced lycopsids (lepidodendrons and sigillarias).

The observation of the fossil site of Joggins, after seeing the fossil sites of the Oka River, brings forth the complementary pictures of terrestrial life in one place and marine life in the other in the late Carboniferous. Similar complementary observations can be made for the terrestrial dinosaurs in the Canadian province of Alberta and the marine dinosaurs in Morden (Manitoba). Land and sea represent two major opposite ecological habitats on the Earth, and the drift of tectonic plates results in changing these habitats throughout the Earth's history.

Chapter 5

Marine Fauna at the End of the Carboniferous Period

5.1. Brachiopods

The largest and best-preserved marine fossils of the Upper Carboniferous are brachiopods. This type of animal (*Brachiopoda*) has a shell but differs significantly from mollusks. It appeared in evolution earlier, and these organisms have a more primitive structure. Most of the brachiopods from the limestones around Kasimov belong to the orders *Productida* and *Spiriferida*.

In the limestones of the Kasimovian Age, the most common are brachiopods of the genera *Marginifera*, *Chonetes*, *Choristites*, and *Neospirifer*; in the Moscow stage, the main genera are also *Marginifera* and *Choristites*. *Marginifera* and *Chonetes* belong to the order *Productida*, and *Choristites* and *Neospirifer* belong to *Spiriferida*. In spiriferids, the shells are mostly ribbed, biconvex in shape, and plano-convex. The skeleton of the outgrowths of the body ("manual apparatus") is in the form of spiral cones. I remember the well-preserved shells of *Choristites mosquensis* near the Vasilyovsky ravine. Productids have shells with a straight locking edge and with small, short "ears."

Brachiopods are characterized by a sedentary lifestyle in shallow seawaters. Their body is covered with a bivalve calcareous shell, which makes them look like bivalve mollusks. The valves of the shells are asymmetrical, the ventral flap is larger than the dorsal and convex, and the dorsal flap is flat and sometimes even concave. The rear edges of the flaps are connected by outgrowths ("lock"). Brachiopods are usually attached to the ground with a stem (leg), and those species that do not have a "leg" burrow into the ground. Brachiopods feed themselves via the filtration of small planktonic organisms and detritus particles suspended in water.

The number of brachiopods has been slowly declining in the last hundred million years, which was the result of the spread and evolution of bivalves, as well as cephalopods, which displaced brachiopods from their habitats. Cephalopod mollusks – ammonites and belemnites – already existed in the

Carboniferous period but became most widespread in the Mesozoic era. Near the village of Alpatyevo, where my grandmother spent her childhood (now it is the Moscow region), there are outcrops of Jurassic deposits on the right bank of the Oka, where belemnites (popularly called devil's fingers) are often found. But in the Kasimov limestones, brachiopods are dominant.

Four major groups (orders) of brachiopods now exist. Fossil brachiopods number up to 30,000 species, grouped into 5,000 genera. Now, about 380 species have survived, united in 100 genera.

5.2. Sea Lilies

Some of the most common fossil remains of the Upper Carboniferous are the stems of sea lilies. Sea lilies are a separate subtype of attached echinoderm with a single class (the other subphylum includes free-moving echinoderms and contains the classes of starfish, sea urchins, ophiures, and holothurians). Primitive sea lilies have been known since the early Paleozoic – from the early Ordovician period, and possibly the Cambrian. *Echmatocrinus* from the Canadian Rocky Mountains from the Burgess Shale Formation is 508 million years old, but its systematic position is not entirely clear. In the Carboniferous period, sea lilies reached their greatest abundance. There were more than five thousand species combined into 11 subclasses, but by the end of the Permian period, most of them became extinct, and now remained or rather appeared from the Triassic period of the Mesozoic era, one subclass of *Articulata*, to which all modern sea lilies belong.

Some limestone beds of the Paleozoic are composed almost entirely of sea lilies. The fossil segments of the stems of sea lilies resemble cogwheels. Now there are about 700 species of sea lilies. Sea lilies are benthic animals with a calyx-shaped bodies, in the center of which there is a mouth, and a corolla of branching rays ("hands") extends upwards. Down from the calyx departs an attachment stem up to one meter long, growing to the seabed and carrying lateral appendages (cirri). At the ends of the cirri, there may be teeth, or "claws," with which stemless lilies are attached to the seabed. Sea lilies retain the orientation of the body characteristic of their ancestors: their mouth is turned upwards, and the dorsal side is turned to the surface of the soil.

5.3. Foraminifera

Foraminifera are single-celled eukaryotic organisms that have an outer skeleton in the form of a shell. They belong to the simplest animals, although this group, according to the classification of Thomas Cavalier-Smith, together with other shelled protozoa – radiolarians, foraminifera, dinoflagellates, diatoms are grouped closer to plants, forming gametes with two flagella (*Bikonta* superkingdom, plants later lost flagella, in contrast to single-flagellated gametes of animals and fungi (*Unikonta* superkingdom) (Cavalier-Smith, 1993).

The internal cavity of foraminifers' shells communicates with the environment through numerous pores as well as through a hole in the shell – the mouth. Through the mouth and the pores in the walls of the shells, the thinnest branching and interconnected reticulopodia (special pseudopods) protrude outward, which serve to move and capture food and form a lattice around the shell, the diameter of which is many times larger than the diameter of the shell. Unicellular algae stick to such a lattice, which foraminifera feed on.

About 4,000 modern species and more than 30,000 species of fossil foraminifera are known. Foraminifera are single-celled organisms, but their shells are quite large. Most foraminifera are less than one millimeter in size, but some fossils reached twenty centimeters in diameter!

Foraminifera species play a major role in the formation of many limestones. Foraminifera built the limestone of the pyramids of Egypt. Currently, coral reef foraminifera produce 43 million tons of calcium carbonate annually. If we take the pyramid of Cheops, then its volume (excluding voids) is 2.5 million cubic meters. Considering that it is built not only of limestone, but also of granite, we take the volume of limestone equal to one million cubic meters and the volume of the average foraminifera – one cubic millimeter. Then, a simple calculation shows that one quadrillion (10^{15}) foraminifera participated in the pyramid limestone production during several million years of the Carboniferous period. Then, during the twenty years in which the pyramid was built, thousands (or tens of thousands) of men in the totalitarian state of Ancient Egypt built a structure of a height of 146 meters, intended for one pharaoh.

The main foraminifera of Kasimovian limestone belong to the genus *Triticites* (order *Fusulinida*). Fusulinids were ubiquitous but became extinct during the Permian-Triassic extinction event 251 million years ago. They were a few millimeters in size, and some reached 2 cm.

The Permian-Triassic extinction was the most devastating for the entire time of the Phanerozoic and perhaps in the last 1-2 billion years. It may have been caused by the fall of an asteroid up to 30 km in diameter in an area that corresponds to modern Western Australia and eastern Antarctica, although this is not proven. Note that the asteroid on the border of the Mesozoic and Cenozoic eras (Cretaceous-Paleogene extinction 66 million years ago), which caused the death of dinosaurs and ammonites, was about 10 km in diameter, therefore, had a mass 25-30 times less). From the Permian-Triassic catastrophe, the Siberian Traps remained – the largest trap province in the world, it is located on the East Siberian platform, and its area is about 2 million square kilometers. Siberian traps poured out on the border of the Paleozoic and Mesozoic (the Permian and the Triassic periods) about 251 million years ago. A significant part of the Siberian Traps is the Putorana Plateau, a highly dissected mountain range located in the northwest of the Central Siberian Plateau, representing a large region of volcanic igneous rock. The volume of erupted melts then ranged from one to five million cubic kilometers.

5.4. Corals

Fossils of corals are common in the Kasimov limestones. They belong to the subclass of six-rayed corals, their two extinct orders – tabulates and rugosa (four-rayed corals). Tabulates are usually smaller than rugosas, and most form colonies (rugosas are often solitary). Since they are six-rayed, the number of vertical calcareous partitions (sclerosepta) in tabulates is a multiple of six, but they are underdeveloped or look like spines; they also have horizontal partitions (tabulae). Tabulates include genera such as *Chaeteles* and *Syringopora*, which are widely represented in the Kasimov limestones and resemble honeycombs.

Rugoses (four-rayed corals) also belong to the six-beam subclass since, at a young age, they have six primary partitions, but then the development of partitions occurs only in four sectors. The name "rugoses" means "wrinkled"; they have characteristic ring wrinkles on the outer surface of the skeleton. These corals have the appearance of a horn extended to the upper end. A characteristic representative of the Kasimovian limestones is the genus *Bothrophyllum*. These fossils are common around Kasimov.

Chapter 6

Role of Land Plants in the Atmospheric O_2 Buildup and CO_2 Depletion

6.1. Gaseous Content of the Atmosphere in the Earth's History

To understand the role of the late Carboniferous biosphere in the establishment of very high atmospheric oxygen and low CO_2 concentrations, we need to analyze the biospheric role of land plants. We will follow the analysis presented in the earlier paper (Igamberdiev and Lea, 2006). When we analyze the periods of highly active biospheric processes fueled by intensive photosynthesis, such as the late Carboniferous, we notice the striking signature of life in the presence of oxygen in the atmosphere at high concentration. This signature is the main indicator of the sustainable non-equilibrium state supported by the activity of the organisms inhabiting the Earth. Among these organisms, the primary role belongs to land plants, which currently contribute to the overall photosynthetic productivity of the planet more than all photosynthetic organisms inhabiting the oceans (Field et al., 1998). Indeed, the atmosphere on Earth is in an extreme state of disequilibrium in which highly reactive gases, such as oxygen and methane, exist at concentrations that are different by many orders of magnitude from the photochemical steady state. Living organisms regulate the composition of the Earth's atmosphere via large biogenic fluxes of gases. These fluxes can be modeled based on feedback regulation of the global environment.

The turnover of CO_2 and O_2 through the biosphere dramatically exceeds the turnover of inorganic geochemical cycles. The annual flux of CO_2 through the biosphere is currently approximately 10% of the atmospheric CO_2, or slightly more than 0.1% of the total carbon in the biosphere. Thus, the turnover time of atmospheric CO_2 is about 10 years, while the turnover of all carbon could be less than one thousand years. Each year, 120 billion tons of carbon are exchanged in each direction between terrestrial ecosystems and the atmosphere; another 90 billion tons is exchanged between the ocean and the atmosphere, while 9 billion tons is emitted by burning fossil fuels and land use (https://earthobservatory.nasa.gov/features/CarbonCycle). For atmospheric

O_2, the turnover time is 4.5 thousand years, while the inorganic cycle is approximately 3.2 million years (Lenton, 1998). Photosynthetic organisms producing O_2 and utilizing CO_2 have played a key role in regulating the gaseous content of the atmosphere, from their first appearance and continuing evolution.

6.2. Concentrations of O_2 During the Phanerozoic Eon

Life originated in an anoxic atmosphere, and the first available O_2 was produced by photosynthesis. The oxygenic photosynthesis of cyanobacteria is a very ancient process and probably appeared in the Archaean era, as early as 3.3-3.5 billion years ago. The presence of O_2 sinks, including photochemical destruction, reduced volcanic and metamorphic gases, and continental weathering, prevented the rise of atmospheric O_2 until the sinks became saturated. After all the inorganic reductants had become exhausted, photosynthesis, whose global overall rate is reflected in organic carbon burial, was needed for oxygen to accumulate in the atmosphere (Bjerrum and Canfield, 2004). The O_2 concentration increased to a few percent (probably ~2%) in the early Proterozoic (Great Oxidation Event) (2.4–2.1 billion years ago), when it decreased again in the middle Proterozoic (Boring Billion) following the episodes of Snowball Earth (Ossa et al., 2022), and further increased in the late Proterozoic. Towards the end of the Proterozoic, its concentration rose to 10%, triggering the Cambrian 'explosion' (the evolutionary diversification of large metazoans) (von Bloh et al., 2003). The apparent diversification corresponding to the Cambrian explosion came from a sudden capacity of metazoans to be calcified and preserved as $CaCO_3$. The molecular clock data indicate that there was a significant diversification accompanying the rise of O_2 preceding the boundary between the Proterozoic and Cambrian (Knoll, 1994). Biological colonization of the land surface started in the late Proterozoic, leading to phosphate and silicate weathering from rocks, decreasing CO_2 concentration, and accumulating calcium carbonate, while the O_2 concentration increased (Lenton and Watson, 2004).

Existing geochemical models indicate an O_2 concentration of 12–15% in early Phanerozoic (Berner and Canfield 1989; Berner 2003), which did not increase much until the emergence and spread of land plants (Lenton, 2001). With the appearance of land plants, which evolved 420 million years ago and were widespread by 370 million years ago, the O_2 concentration increased, reaching a peak near 300 million years ago (late Carboniferous). Giant dragonflies, charcoal deposits, and indications of intensive fires provide evidence for concentrations of O_2 higher than present. Berner et al. (2003) have suggested a maximum of 35% O_2 in the late Carboniferous. After this period, the concentration of O_2 decreased and was relatively stable, falling during the Triassic and Jurassic below 20% (with a corresponding disappearance of fires and charcoal burial) and increasing in the Cretaceous (150 million years ago) to the present-day value or slightly higher (up to 25-27%) (Lenton 2001; Bergman et al. 2004). At the O_2 concentration of 30-35% in the late Carboniferous, forest fires were even more common than now; however, there would be no widespread burning of forests following a single lightning strike, at moisture contents common to living plants (Wildman et al., 2004). Times of high O_2 coincide with observations of fire-resistant plant morphology, large insects, and high concentrations of fossil charcoal (Wildman et al. 2004).

The progressive oxygenation of the Earth's atmosphere was crucial for life evolution. Redox proxy data indicate the deep oceans were oxygenated at the end of the Silurian Period (435 million years ago), and fossil charcoal indicates the O_2 concentration of more than 15-17% in the early Devonian Period (420-400 million years ago). The earliest plants, which colonized the land surface starting from the Ordovician Period (bryophytes started to colonize land approximately 470 million years ago), became the cause of the mid-Paleozoic oxygenation event (Lenton et al., 2016). Early plants selectively increased the flux of phosphorus weathered from rocks, which stimulated the life and reproduction of marine animals that have a much higher phosphorus to carbon ratio as compared to plants.

It is becoming evident that vegetation cover achieved about one-third of today's global terrestrial net primary productivity by the late Silurian (445 million years ago) (Lenton et al., 2016). In the Carboniferous period, the productivity could be comparable to the current values. At that time, the growth and transpiration rates of the key Carboniferous plants approached the values found in some angiosperms and had a very high potential for physiological forcing of climate through vegetation-climate feedbacks (Wilson et al., 2017). In the upper latitudes during the periods of cooling, the productivity was limited by the parameters of freezing tolerance of the Carboniferous plants (Matthaeus et al., 2021).

6.3. Concentrations of CO_2 During the Phanerozoic Eon and Limestone Deposition

Modern geochemical modeling based on the distribution of ^{13}C and other isotopes shows that before the fall in CO_2 caused by the appearance of land plants (initially of the bryophytes in the Ordovician), the early Phanerozoic concentration of CO_2 was 0.4–0.5%. The CO_2 concentration dropped first in the Ordovician when mosses colonized land (Donoghue et al., 2021), then further during the Devonian and the Carboniferous (coinciding with the highest O_2 concentration in the late Carboniferous) (Bergman et al., 2004). It was higher during the Mesozoic era, probably occasionally reaching 1500-2000 ppm (0.15-0.2%). Later in the Cenozoic era, a gradual decrease of CO_2 took place, and for nearly last 50 million years, the concentration was not much higher than the present preindustrial level of 300 ppm (Royer et al., 2001). Detailed analyses of the air bubbles in cores of Antarctic ice have indicated that the CO_2 concentration oscillated between 180 and 280 (maximum 300) ppm during the last 420 thousand years (Di Michele et al., 2001). In the last 740 thousand years, eight big glacial cycles occurred (EPICA community members, 2004), with the oscillations of a 100 thousand years period, with a slow fall and a rapid increase (less than 10 thousand years). Ten to fifteen smaller oscillations with an amplitude of 10–20 ppm were observed within every large oscillation. Likely, a similar occurrence of oscillating CO_2 with corresponding glaciations and deglaciations took place around the Carboniferous-Permian boundary.

During this time, the shelled animals inhabiting seas and oceans were very abundant, which led to the deposition of large amounts of calcium carbonate in the form of limestone and chalk. Under high CO_2 conditions, foraminifera had major limitations in building shells, repairing spines, and were physiologically stressed, while at low CO_2, they had optimal conditions for life and reproduction (Davis et al., 2017). The same applies to other shelled animals. The geological record of ocean acidification and its effect on shelled organisms shows that they were flourishing in the periods of decreased atmospheric CO_2, when calcium carbonate is more stable at higher pH values (Hönisch et al., 2012). While land plant life had limitations at low CO_2 due to high photorespiration, the concentration of bicarbonate and carbonate in water always remained sufficiently high to build invertebrate skeletons and shells. The formation of limestone from seawater occurs mainly via its biogenic precipitation by lime-secreting organisms, and the most powerful among them are foraminifera, which were abundant from the end of the Proterozoic eon

(Pawlowski et al., 2003). The depletion of CO_2 by plants in the atmosphere was accompanied by active deposition of limestone by marine organisms, which formed the landscape patterns in different regions of the Earth.

The deviations in the temperature of the Earth's atmosphere can be correlated with the CO_2 concentration, and even the equation for such correlation has been derived (Berner and Kothavala, 2001). However, the temperature on the Earth depends not only on CO_2, thus all these estimations remain approximate. The data based on foraminifer shells extracted from impermeable clay-rich sediments show that the distribution and replacement of the species of these temperature-sensitive organisms in clay-rich sediments during the late Cretaceous and Eocene epochs is in agreement with the variations in CO_2 concentrations (Pearson et al., 2001). Correlations between temperature and CO_2 concentration during the last 400 thousand years are very precise (Cuffey and Vimeux, 2001). Some deviations in earlier periods have been demonstrated, e.g., in the Miocene, where the medium temperatures were 6°C higher than some authors predicted from a CO_2 concentration of 250–290 ppm, which could be explained, in particular, by the presence of other greenhouse gases, e.g., methane (Zachos et al., 2001).

6.4. Land Plant Photosynthesis and the O_2/CO_2 Balance

In the photosynthesis process, CO_2 is fixed via the Calvin–Benson cycle, in which ribulose-1,5-bisphosphate carboxylase/oxygenase (Rubisco) is the primary CO_2 assimilatory enzyme in C_3 plants. During evolution, the enzyme has preserved a capacity to use O_2 as a substrate, which in an atmosphere of high O_2 and low CO_2, makes CO_2 fixation less efficient due to the photorespiratory process, starting with the oxygenation reaction of Rubisco. This reaction becomes physiologically important in the atmosphere with low CO_2 and high O_2, causing the photorespiration process. Metabolic pathways have evolved to overcome the loss of CO_2 during photorespiration, either via a CO_2 concentrating mechanism (CCM) in algae or via the C_4 pathway of photosynthesis in some advanced higher plants (and also in some diatoms, which are unicellular algae). In C_4 photosynthesis, primary CO_2 fixation is carried out by another enzyme (phosphoenolpyruvate carboxylase), which has no oxygenase reaction, and CO_2 is delivered to the Calvin–Benson cycle at a higher concentration in a specific compartment, or, in rare cases, in the other part of the same cell as in the single-cell C_4 photosynthesis. Below, we will discuss the biospheric consequences of the development of photosynthesis on Earth and its role in preserving the O_2/CO_2 balance.

The Rubisco enzyme has maintained during evolution the affinity for oxygen. The Rubisco specificity factor defines the preference of Rubisco for CO_2 compared to O_2 (Laing et al., 1974). The value of this factor is ~100 at 25°C in the angiosperm C_3 species, while it is lower in C_4 plants, in conifers, and ferns, and it is significantly lower in some green algae (Jordan and Ogren, 1983). The O_2/CO_2 ratio in the atmosphere is about 36 times higher than in solution at 25°C because of the higher solubility of CO_2 compared to O_2. This means that the value for the specificity factor of Rubisco is ~3600 in relation to the atmospheric O_2/CO_2 ratio. This indicates that the Rubisco enzyme will have a 3600 times higher affinity for CO_2 than for O_2 in the atmosphere, provided there are no limitation effects of stomatal conductance. In the real situation in plants, we have to consider also the concentration of CO_2 near the sites of carboxylation, which can be 50% less than that in air (Evans and von Caemmerer, 1996), or even less for xeromorphic plants (Di Marco et al., 1990), for estimating the specificity of Rubisco in the sur- rounding atmosphere.

6.5. Difference in Photosynthesis of Land Plants and Algae

Only the unusual Rubisco enzyme from certain thermophilic red algae (cyanidiophytes) has very high carboxylation to oxygenation (Vc/Vo) ratios (Uemura et al., 1997), which can be explained by a higher affinity for CO_2, rather than a lower affinity for oxygen. However, the value of the Rubisco specificity factor in cyanidiophytes is much less extreme when the in vitro assays are conducted at the growth temperatures of the organisms, and the species of red algae living at moderate temperatures have Rubisco with much less variation in its specificity to CO_2 and O_2 (Uemura et al., 1997). The highest value of the Rubisco specificity factor (310 at 90°C) was reported for the enzyme of the hyperthermophilic archaeon, *Pyrococcus kodakaraensis* (Ezaki et al., 1999). Tortell (2000) has shown that plotting the Rubisco specificity factor for a range of algae against the date of evolution gives a good correlation, showing its increase during evolution, and the red algae really stand out as having different values. However, the question of why this high affinity enzyme was not positively selected for during the evolution of other groups leading to the appearance of land plants remains open. The Rubisco from red algae can be expressed abundantly in transgenic higher plant chloroplasts but is not assembled to form an active enzyme (Whitney et al., 2001). The expression of red algal Rubisco with its cognate activase could potentially overcome these problems (Gunn et al., 2020).

The specificity factor of Rubisco can impose limits on the CO_2 and O_2 concentrations in the atmosphere. Without special concentrating mechanisms for CO_2, plants can only exist in a range of specified concentrations of O_2 and CO_2. Above a certain concentration of O_2 and below a certain concentration of CO_2 (compensation points), the oxygenase reaction of Rubisco will dissipate more carbon than is fixed in the carboxylase reaction (Tolbert et al., 1995). Even small changes in the concentrations of O_2 and CO_2 in the atmosphere can lead to drastic changes in metabolism, favoring either reductive or oxidative reactions. At low CO_2 levels, the CO_2 concentrating mechanisms (CCM) that evolved in algae become important (Badger and Price, 2003). The CCMs evolved as a response to the decrease in CO_2 concentration during the Paleozoic. The CCMs use energy for the active transport of protons, CO_2, or bicarbonate. In algae, high rates of photosynthetic carbon assimilation occur even at CO_2 concentrations as low as 5 ppm, because of an effective CCM based on carbonic anhydrase (Raven, 2003). This mechanism is induced by low CO_2 concentrations, being preceded by the photorespiratory release of glycolate and, to a lesser extent, of glyoxylate, glycine, and CO_2. Photorespiratory peroxisomes with the same enzyme content as in land plants appeared only in their immediate ancestors, the Charophyceae algae (Karol et al. 2001). The absence, or very low activity, of a CCM and active peroxisomal metabolism, together with a high glycine decarboxylase capacity in mitochondria, has made the photorespiratory process very intensive in land plants.

The role of phytoplankton in O_2 release was significant from the time in the Proterozoic eon, when all inorganic reductants had been exhausted and organic carbon burial began to take place (Bjerrum and Canfield, 2004), while CO_2 depletion was mostly due to the activity of land plants via the processes of weathering (in soil formation) and active photosynthesis. When the CO_2 concentration in the atmosphere decreased to a minimum corresponding to the ecological compensation point in land plants, the productivity of land plants diminished. The ecological compensation point is a CO_2 concentration (on average 150–180 ppm at 21% O_2), below which plants are unable to complete their lifecycles (Sage and Coleman, 2001). During the glacial periods, when the atmospheric CO_2 concentration drops to such low levels, temperature gradients drive oceanic circulation, causing a greater supply of oxygen to deep waters (Berner, 2003). This would result in an increase in atmospheric CO_2

via inhibition of the formation of anoxic waters in the bottom layers and a reduction of organic burial.

6.6. Respiration and Photorespiration

The reduction level of higher plants, reflecting the reduction state of reaction centers in chloroplasts and pyridine nucleotides in all subcellular compartments, primarily in mitochondria, is strongly affected by the O_2/CO_2 ratio. At low CO_2 and high O_2, the over-reduction of chloroplast reaction centers occurs, due to the low rate of CO_2 fixation in the Calvin-Benson cycle. Under these conditions, photorespiratory ammonia assimilation, together with other energy-utilizing and/or wasting processes, consumes the excess reduction power in the chloroplasts (Igamberdiev and Lea, 2002). At the same time, in mitochondria, photorespiration and respiration increase the reduction level, moderated by the operation of non-coupled pathways. This results in the incomplete oxidation of glycolytic and photorespiratory substrates, leading correspondingly to the efflux of citrate from mitochondria (Igamberdiev and Bykova, 2023) and the suppression of glycine oxidation resulting in oxalate formation in peroxisomes (Igamberdiev and Lea, 2002), both compounds being further excreted into the soil.

Photorespiration is essential even at much higher atmospheric CO_2 concentrations than at present. High CO_2 concentrations result in an elevation of temperature, which increases photorespiration and dark respiration rates more than the rate of assimilatory carboxylation (Brooks and Farquhar, 1985). The development of large leaves in the course of evolution may have prevented overheating due to higher transpiration rates (Beerling et al., 2001). Based on the temperature dependence of the oxygenase to carboxylase ratio of Rubisco and the estimated deviations of surface temperature from the mean value, we can calculate the oxygenase to carboxylase ratio during the Phanerozoic eon according to the estimates of CO_2 and O_2 in the atmosphere. The ratio of the photorespiratory rate to the photosynthetic assimilation rate is half of the oxygenase to carboxylase ratio (Sharkey, 1988). Paleozoic photorespiratory rates could be slightly higher as the carboxylase function of Rubisco was lower by 10–20% (Bird et al., 1982), but these rates could also respond to a lower brightness of the sun in the Paleozoic (by 5%), which would result in lower temperatures.

6.7. Land Plant Photosynthesis and Photorespiration in the Regulation of O_2 and CO_2

Tolbert et al. (1995) developed the concept that CO_2 depletion and O_2 release caused by photosynthesis are counterbalanced by CO_2 release and O_2 uptake during photorespiration. In an experiment with tobacco and spinach in closed chambers, the assimilation of CO_2 led to its relative exhaustion down to low concentrations and to an increase of O_2 with the establishment of an equilibrium O_2/CO_2 ratio. In other experiments, it has been shown that plants can complete their lifecycles at the concentrations of 35% O_2 and 350 ppm CO_2, which were established in the late Carboniferous and early Permian (Beerling and Berner, 2000). Even lower CO_2 concentrations at that time may not completely limit plant productivity. Land plant photosynthesis not only contributes to the establishment of the O_2/CO_2 ratio, but it also mediates the processes of CO_2 depletion in weathering and in the operation of the ocean sink, so this depletion never falls below the ecological CO_2 compensation point, i.e., the feedback mechanisms switch upon depletion, resulting in a rise of CO_2. Thus, the biospheric equilibrium of CO_2 and O_2 concentration works with feedback mechanisms that may be responsible for the oscillatory regime, together with the synchronization of these oscillations with the cycles of solar activity.

6.8. The Role of Weathering by Land Plants in CO_2 Depletion in the Atmosphere

The removal of CO_2 from the biosphere occurs not only directly in the photosynthesis process but also via the weathering of silicates and formation of carbonates, during root activity and soil formation (Berner, 1997). The plant activity in the process of weathering is apparently more important than temperature and rainfall. The weathering process is directly connected with the photosynthetic activity of land plants and a requirement for phosphate, being caused mainly by the excretion of citrate and oxalate from roots (Diatloff et al., 2004). Oxalate is formed, in particular, as a side product of the glycolate oxidase reaction in leaves, when the reduction level in mitochondria established during high rates of photorespiration suppresses the oxidation of glycine (Igamberdiev and Lea, 2002). Citrate is also formed when the reduction level in mitochondria increases, due to switching from the complete to the partial TCA cycle (Igamberdiev and Bykova, 2023). Organic acids can be transported to roots, or alternatively, citrate can be formed in roots due to the high activities of citrate synthase (Kihara et al., 2003). At the pH of phloem sap, oxalate salts are transported together with citrate (which constitutes one-third of all acids transported) and malate. The transport through the phloem flow together with the formation of organic acids in the rhizosphere constitutes an effective mechanism of organic acid excretion by the root system (Jones, 1998).

In the weathering process, citrate and oxalate release phosphate and other mineral compounds during soil formation. The release of phosphate leads to the eutrophication of the ocean and to the increase of photosynthetic productivity by algae, as well as to the substitution of silicates by carbonates, both processes consuming atmospheric CO_2 (Lenton, 2001). The CO_2 depletion by weathering was more intensive than the release of CO_2 to the atmosphere due to geological and biological activities, including fires and respiration in extended periods, except during the short times of volcanic eruptions, etc. This led to a decrease in the atmospheric concentration of CO_2 up to the time of glaciation, when the CO_2 production rate increased. Glaciation strongly decreased the process of weathering due to both lower rates of metabolism in the cooler and drier climate and less plant cover, and the CO_2 concentration began to rise mostly via negative feedback on oceanic phytoplankton, which results in the release of CO_2 more intensively as the CO_2 concentration falls (Raven and Falkowski, 1999). At low CO_2 concentration, fast-growing grasslands are predominant, while when the CO_2 increases, slower-growing trees become widespread (Bond et al., 2003), contributing to different rates of weathering and CO_2 production.

Vascular plants amplify the rate of weathering by about an order of magnitude relative to lichens and mosses (Lenton, 2001). Photosynthetic uptake of CO_2 (and release of O_2) is counterbalanced by respiration (and photorespiration in land plants). The photosynthesis process will deplete CO_2 if the burial of organic carbon takes place, which was higher at the Permian–Carboniferous boundary (Lenton, 2001). The processes of organic carbon weathering and degassing are usually estimated to be approximately the same rate as photosynthetic CO_2 assimilation, but this is a matter of debate, especially for earlier times, e.g., for late Devonian and Carboniferous (Lenton, 2001). During the Mesozoic era, the weathering process was probably less intensive because of the type of flora (gymnosperm) that existed (Moulton et al. 2000). This could correspond to higher concentrations of CO_2. The distribution of angiosperms in the Cretaceous could cause a further increase in weathering and a decrease in CO_2 in the atmosphere.

It is difficult to estimate the relative input of Tolbert's factor (photosynthesis/photorespiration balance) and Berner's factor (silicate weathering) on CO_2 depletion. The burial of carbon in the Carboniferous was a significant factor in CO_2 depletion. While the weathering process likely initiates the CO_2 depletion, at the approximation of the achievement of the CO_2 minimum and O_2 maximum, the depletion of CO_2 is controlled mainly by photosynthetic assimilation. Near this critical point, plants at lower latitudes (at higher temperatures) will produce CO_2, while at higher latitudes (lower temperatures), they will consume it, but this, in turn, is limited by the decrease in temperature. Together with the uptake of CO_2 by the ocean, the processes of CO_2 assimilation and CO_2 release would have contributed to the glacial-interglacial oscillations, which are also synchronized with solar activity (the Milankovitch cycles). The global effects of the biosphere on the balance of CO_2 in the atmosphere include photosynthetic CO_2 assimilation, photorespiration, weathering, temperature, etc. For the modeling of these events, it is necessary to know the rate constants of all the processes and their dependence on temperature and CO_2 concentrations. The removal of CO_2 by weathering and by direct photosynthetic activity can be estimated from the relative flux intensities of these processes. At present, this information remains incomplete.

6.9. The O_2/CO_2 Compensation Ratio

To understand the role of land plants in the equilibration of O_2 and CO_2 concentrations in the biosphere, we need to discuss the concept of the compensation point (Γ). The CO_2 compensation point is determined as the concentration of CO_2 when the rate of photosynthetic CO_2 assimilation is equal to the rate of CO_2 release by respiration and photorespiration. If we ignore leaf respiration, which is relatively low in the light and probably does not exceed 5% of the rate of assimilation (Atkin et al., 2000), we get gamma star (Γ^*), which is the compensation point that takes into consideration only photorespiration. Definitions of the CO_2 compensation point are based on a constant O_2 concentration, as it has been shown that the dependence of Γ on O_2 is linear (Farquhar et al., 1980). The value of Γ for C_3 plants at 21% O_2 is about 50 ppm CO_2 at 25°C. However, the real ecological compensation point is higher (because of the necessity of photosynthetic and respiration costs for maintenance, growth, and productivity), being in general about 180 ppm, depending on temperature, irradiance, humidity, and other factors (Sage and Coleman, 2000). Respiration by plant and non-plant organisms may contribute to the value of the ecological compensation point, or at least have some feedback impact on the decrease of atmospheric CO_2. Below some lesser CO_2 value than the ecological compensation point (at the O_2 value of 21%), plants are unable to complete their lifecycles (Sage and Coleman, 2000). Lack of carbon reduces the capacity of plants to assimilate nutrients, particularly nitrogen (Andrews et al., 2004).

Tolbert et al. (1995) introduced the definition of the O_2 compensation point, which means the concentration of O_2 at a constant concentration of CO_2, when the assimilation and the release of CO_2 are equal. Combining the definitions for CO_2 and O_2 compensation points, we get the definition of the O_2/CO_2 compensation ratio, which represents the limit of the ratio of O_2 concentration to CO_2 concentration. For Γ equal to 50 ppm CO_2, the O_2/CO_2 compensation ratio is 4200. However, the ecological O_2/CO_2 compensation ratio (considering the ecological compensation point of 180 ppm) is about 1200. Probably this O_2/CO_2 compensation ratio occurred in the Carboniferous–Permian time and the Pleistocene glacial times. The values of the compensation point and of the compensation ratio also depend on temperature, which changed during the Phanerozoic eon, and should be considered.

6.10. Correlation of O_2 and CO_2 in the Atmosphere with the Evolutionary Process

If the diversity of land plants expressed as the number of land plant families is compared with the concentrations of O_2 or CO_2 in the atmosphere, there is no obvious correlation between the diversity and concentration of either O_2 or CO_2. However, there is a surprisingly good correlation between the O_2/CO_2 ratio in the atmosphere and biodiversity (Igamberdiev and Lea, 2006). The correlation is even better if biodiversity is plotted against the effective O_2/CO_2 ratio (the ratio of oxygenase to carboxylase rates of Rubisco, Vo/Vc), calculated based on the effect of temperature on the compensation point. The increase in this ratio correlates well with diversity, while a decrease is accompanied by the extinction of land plant families. A similar correlation has been observed with the number of animal species (Rothman, 2001). Elevation of CO_2 under constant O_2, i.e., a decrease of the O_2/CO_2 ratio, has previously been shown to reduce biodiversity in land communities (Zavaleta et al. 2003).

Only some parameters of plants are correlated directly with O_2 but not with CO_2, e.g., the replacement of woodiness by herbaceousness in evolution and the woodiness indices of plants (Gottlieb and Borin, 1998). The evolution of secondary metabolites was triggered by O_2, providing the incorporation of oxygen atoms into a series of molecular species (Gottlieb and Borin, 1998). Contrary to land plants, the diversity of marine fauna follows directly the CO_2 concentration and corresponding temperature changes (Cornette et al., 2002). This should be considered for the evaluation of the rates of evolution of marine species in the late Carboniferous.

The increase in biodiversity during the late Devonian and early Carboniferous corresponded to the increase in the O_2/CO_2 ratio, peaking at the late Carboniferous–Permian boundary. This was the period of the highest diversity of land plants before the late Mesozoic. The diversity could be even higher, since, e.g., angiosperms became distributed widely only in the Cretaceous, but comparisons of gene sequences showed that the differentiation between gymnosperms and angiosperms arose at the Carboniferous–Permian boundary (Savard et al., 1994). A Carboniferous–Permian high O_2 episode might have triggered this split between major plant groups. Recent data have indicated the existence of oscillations of CO_2 (and consequently temperature) of a 100 thousand years period (Sigman and Boyle, 2001), possibly with minimum CO_2 concentrations (corresponding to glaciations) close to that in the Pleistocene oscillations (Tajika, 2003), which is reflected in the estimations of CO_2 concentrations in the paper of Richey et al. (2020) (see Figure 6). There is also some indication of the appearance of C_4-like plants in the late Carboniferous period based on the $\delta^{13}C$ value (19‰) of some fossils (Jones, 1994), but this requires further substantiation (Cowling, 2013).

The appearance and development of stomata significantly influenced the evolution of land plants. The stomata evolved in the Silurian, and the stomatal index increased as carbon dioxide decreased from the Silurian to the Carboniferous (Royer et al., 2001). The decrease in CO_2 and increase in O_2 led to the changes resulting in maximizing CO_2 diffusion into the leaf (Beerling and Woodward, 1997), thereby raising intercellular CO_2 concentrations and reducing CO_2 evolution by photorespiration. In addition, high transpiration rates through stomata prevented the overheating of leaves, allowing the evolution of larger leaves (Beerling et al., 2001).

The decrease in O_2 and the increase in CO_2 during the Permian period were accompanied by a marked decrease in the diversity of land plants. The catastrophic event of the Permian–Triassic boundary, causing the greatest extinction in the history of Phanerozoic, completed this process, destroying 90% of the marine animal species and 70% of the plant species (Rothman, 2001). It could be caused by the major asteroid impact, although this is not established with certainty. While all impact events, after a short period of cooling, led to global warming and a rise in atmospheric CO_2, this rise probably determined the relatively low diversity of the Triassic period. The determination of atmospheric CO_2 concentrations based on stomatal characteristics shows the periods of instant CO_2 increase caused by the impacts of asteroids, more clearly than the isotopic data (Retallack, 2001). This can be seen at the boundaries of Permian–Triassic, Triassic–Jurassic, and Cretaceous–Palaeogene.

Plant metabolism can be more reductive or more oxidative, depending on the O_2 and CO_2 concentrations in the atmosphere (Cen et al., 2001). The increased O_2/CO_2 ratio is more striking at higher temperatures, providing a higher Rubisco Vo/Vc ratio, and consequently a higher compensation O_2/CO_2 ratio. This determines a higher biological diversity in tropical areas, followed by the spreading of plants from the tropical areas towards moderate and subpolar latitudes (Meyen, 1987). In addition, the rate of the evolutionary process triggered by an increase in the O_2/CO_2 ratio is mediated by the level of reactive oxygen species. At higher altitudes, where active oxygen (ozone) is at a high concentration and ultraviolet radiation is strong at lower partial pressures of atmospheric gases, the rate of evolution is higher. The reduction level in plant cells strongly depends on the O_2/CO_2 ratio (Cen et al., 2001). Low CO_2 greatly enhances plant stress symptoms, while high CO_2 alleviates these effects (Cowling and Sage 1998). At high O_2 to CO_2 ratios, oxygen can easily be converted to toxic superoxide and hydrogen peroxide, thus causing gene mutations (Raven, 1991).

An increase in angiosperm biodiversity during the Cenozoic era caused, for the second time during the Phanerozoic eon, the global long-term depletion of CO_2 in the atmosphere down to the ecological compensation point. The appearance of most C_4 plant species is connected with the late Miocene (8–10 Ma); however, in some special areas, they appeared earlier and then spread over the Earth. The C_4 pathway independently evolved at least 62 times in 19 families of angiosperms, dating back 25–32 million years ago (Lyu et al., 2024). The C_4 plants appeared in tropical areas where the compensation ratio is lower, and thus photorespiration is higher. However, C_4 plants have a limited capacity for spreading into colder areas because of the additional energy needed to provide the operation of the C_4 cycle (Sage, 2004).

The Miocene epoch of the Neogene period of the Cenozoic era was characterized by quite low CO_2 concentrations (250–290 ppm) decoupled from temperature (higher by +6°C than calculated based solely on the CO_2 concentration), probably because of outbursts of greenhouse methane (Zachos et al., 2001). This high temperature and a high O_2/CO_2 ratio caused a climate favorable for the C_4 type of photosynthetic metabolism. The C_4 plants are adapted to low CO_2 concentrations and warm climates, and their photosynthetic metabolism is considered an efficient CO_2 pump (von Caemmerer and Furbank, 2003). They, however, can also prosper at elevated CO_2 concentrations, particularly at elevated temperatures and in arid conditions (Sage and Kubien, 2003).

Thus, there is a good indication that the appearance of new genetic material is correlated with the periods when the O_2/CO_2 ratio is maximal. During the last million years, this was observed during the glaciation periods. Tropical/subtropical areas and mountain regions with a higher degree of oxygen effects were the centers for the origin of cultivated plants (Vavilov, 1926). Recessive genes then drifted to the peripheral regions of the species' distribution. Cultivated plants are characterized by a channeling of metabolism towards a higher productivity of certain storage tissues. The genetic diversity was likely built up in tropical areas during the glaciation periods, while the spreading occurred during the interglacial periods. Thus, the origin of agriculture was linked to the transition from glaciation to higher CO_2 concentrations, when agriculture could be effective (Igamberdiev, 2025).

6.11. Greenhouse Bursts and Their Consequences

The rate of increase of CO_2 concentration in the atmosphere was rapid (several-fold in short periods) on several occasions during the Phanerozoic, mostly after impacts of large meteorites (post-apocalyptic greenhouse effects). After these events, CO_2 was established at a higher concentration. At the Cretaceous–Paleogene boundary, there was an increase in CO_2 from 350–500 ppm to 2300 ppm within less than 10,000 years according to the data on stomatal densities (Retallack, 2001). The CO_2 concentration then decreased to the initial value during hundreds of thousands of years (Beerling et al., 2002). The warming impaired leaf photosynthetic function and severely reduced carbon uptake (McElwain et al., 1999). During the Cretaceous–Paleogene boundary, carbon isotopic recovery (return to the ^{13}C values before the asteroid impact) took 100–200 thousand years (Ahrens and Jahren, 2000;

Beerling et al., 2001). This decrease in CO_2 concentration was due to the unique role of land plants in controlling the O_2/CO_2 ratio. The terrestrial ecosystems were recovering ahead of marine production (Beerling et al., 2001), however, this process was relatively slow compared to the fast greenhouse bursts.

Higher CO_2 concentrations, which persisted for long periods of millions of years, may have resulted either from an unusual solar activity or from the specific properties of the flora with a low weathering rate and slow metabolism (possibly gymnosperms in the Mesozoic era). According to the above considerations, we have assumed a major role of photorespiration in maintaining a climate suitable for biospheric development. During the last one million years, oscillations of CO_2 concentration near the ecological compensation point were accompanied by corresponding climate changes. They could be explained by a self-regulatory role of the biosphere (a decrease of weathering at the lowest CO_2 concentrations leading to a CO_2 increase), synchronized with solar activity changes due to the changes in precession of the Earth's orbit described as the Milankovitch cycles (Rial, 2004). The period of these oscillations was around a hundred thousand years in the last million years, and the amplitude was between 180 and 280 ppm, i.e., by about 60% (Sigman and Boyle 2001). In the early Pleistocene epoch, the oscillations were faster with lower amplitude. Now the indications are that these oscillations will cease due to the anthropogenic release of CO_2 (Falkowski et al., 2000). A doubling of the CO_2 concentration is predicted in 100 years or even less, and no natural process can effectively cope with this increase. The biosphere may return to the pre-Devonian state, in which the high CO_2 concentration was not fully regulated by the biosphere. The current increase in CO_2 concentration, caused both by anthropogenic burning of fossil fuel and by unusual solar activity, is unique, and a similar CO_2 increase did not occur from the early Cenozoic. If we consider the predicted CO_2 concentration increase later than the year 2100, the possible effect may be compared even with the early Devonian period. This situation becomes beyond human control.

6.12. Concluding Remarks

Before the emergence of land plants, the photosynthetic activity of marine and freshwater organisms was insufficient to deplete CO_2 from the atmosphere, the concentration of which was more than an order of magnitude higher than present. With the appearance of land plants, the exudation of organic acids by roots, following respiratory and photorespiratory metabolism, led to phosphate weathering from rocks, thus increasing aquatic productivity. Weathering also replaced silicates with carbonates, thus decreasing the atmospheric CO_2 concentration. Because of both intensive photosynthesis and weathering, CO_2 was depleted from the atmosphere down to low values approaching the compensation point of land plants. During the same period, the atmospheric O_2 concentration increased to maximum levels about 300 million years ago (Permo-Carboniferous boundary), establishing an O_2/CO_2 ratio above 1000. At this point, land plant productivity and weathering strongly decreased, exerting negative feedback on aquatic productivity. Increased CO_2 concentrations were triggered by asteroid impacts and volcanic activity, and in the Mesozoic era could be related to the gymnosperm flora with lower metabolic and weathering rates.

A high O_2/CO_2 ratio is metabolically linked to the formation of citrate and oxalate, the main factors causing weathering, and to the production of reactive oxygen species, which trigger mutations and stimulate the evolution of land plants. The development of angiosperms resulted in a decrease in CO_2 concentration during the Cenozoic era, which finally led to the glacial-interglacial oscillations in the Pleistocene epoch. Photorespiration, the rate of which is directly related to the O_2/CO_2 ratio, due to the dual function of Rubisco, may be an important mechanism in maintaining the limits of O_2 and CO_2 concentrations by restricting land plant productivity and weathering. Photorespiration has played an important role since the appearance of land plants in the regulation of the O_2 and CO_2 concentrations in the atmosphere. The limits of photosynthetic/photorespiratory parameters based on Rubisco kinetics, determined limits of variation in O_2 and CO_2 concentrations (the O_2/CO_2 ratio). The established atmospheric O_2/CO_2 ratio was coincident with the rates of evolution of land plants.

The abundance of animals with hard skeletons inhabiting seas and oceans, which we observe in the Kasimovian limestone, is linked to the photosynthetic activity of land plants. Under high CO_2 conditions, invertebrates have major limitations in building shells, while at low CO_2, they have optimal conditions for life and reproduction. Shelled organisms flourished in the periods of decreased atmospheric CO_2, when calcium carbonate is more stable at higher pH values. Low CO_2 limited the growth of land plants due to photorespiration, but the concentration of bicarbonate and carbonate ions in water remained sufficiently high for the maintenance of the productivity of algae possessing carbon concentration mechanism, and for building invertebrate shells and vertebrate skeletons. This led to the deposition of large amounts of calcium carbonate during the ages of the Upper Carboniferous period.

Chapter 7

Events of the Quaternary Period and the Role of Carboniferous Limestone

7.1. Glacial-Interglacial Oscillations in the Quaternary Period

From the Carboniferous times, we will jump over 300 million years to the current geological period as the geological history near Kasimov does not bring records of the epochs between the Carboniferous and the Quaternary periods, when the climate was warmer than at the end of Carboniferous and than it is now. During the Quaternary period, which began just over two and a half million years ago and in which the evolution of the genus *Homo* took place, the Earth turned to an icehouse climate. The Earth's climate was constantly changing, with the alternation of long glaciations and shorter warm interglacial times. The largest climate fluctuations lasted about 100,000 years during the second half of the period (just over a million years) and 41,000 years during the first half of the period. Within these time intervals, shorter fluctuations occurred, usually with smaller amplitudes of temperature changes. The cooling was usually long and gradual, while the exit from the ice age was faster.

Over the past half a million years (Middle and Upper Quaternary), there have been four large glaciations. On the territory of the East European Plain, they are called the Oka glaciation (500-420 thousand years ago), the Dnieper glaciation (300-250 thousand years ago), the Moscow glaciation (180-130 thousand years ago), and the last Valdai glaciation (70-12 thousand years ago). The Valdai glaciation corresponds to the Vistula glaciation in Central Europe, the Würm glaciation in the Alps, and the Wisconsin glaciation in North America. The Moscow glaciation corresponds to the Warta glaciation in Central Europe. The Dnieper glaciation was at the same time as the Saal glaciation in Central Europe, the Riess glaciation in the Alps, and the Illinois glaciation in America. The Oka glaciation corresponded to the Elster, Mindel, and Kansas glaciations, respectively. Shorter intervals between glaciations correspond to the interglacial epochs of Likhvin, Odintsovo, Mikulin (Eemian), and Mologo-Sheksna. The latter (Mologo-Sheksna) corresponds to the beginning of the retreat from the last glacial (the Valdai glaciation in Eastern Europe) period, which was interrupted by a sharp cooling for a thousand years, called the Younger Dryas (12,900-11,700 thousand years ago), probably caused by the fall of a meteorite with a diameter of a third of a kilometer in North America. The beginning of the Holocene epoch (11.7 thousand years ago) corresponds to the modern warm period, during which human civilization developed.

Figure 8 shows the fluctuations in carbon dioxide and temperature over the past 420,000 years, recreated from the analysis of the Antarctic and Greenland ice bubbles. During this time, carbon dioxide and temperature correlated well, but changes in carbon dioxide were delayed by several hundred years. At present, the carbon dioxide content due to anthropogenic activities has reached 425 parts per million (December 2024), and the temperature has increased by 1.5 degrees Celsius compared to preindustrial times. While greenhouse gas emissions currently determine the rise in temperature, in prehistoric times, the temperature changes caused by the precessions of Earth's orbit (Milankovich cycles) caused the changes in CO_2 and methane concentrations in the atmosphere determined by temperature-dependent changes in the solubility of greenhouse gases in oceans and seas.

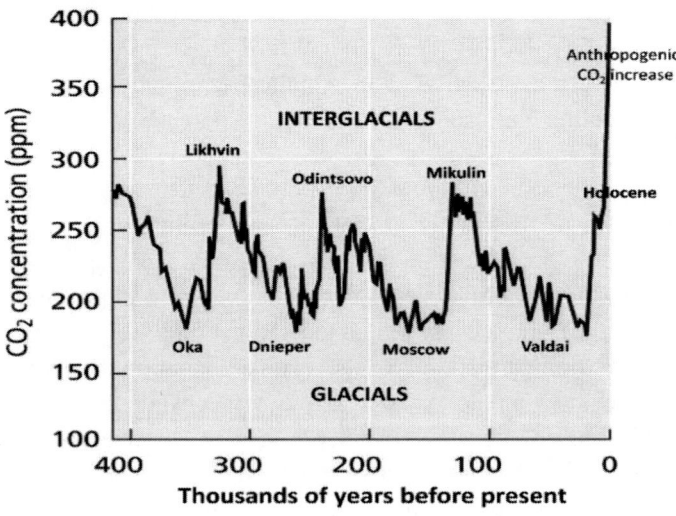

Figure 8. Changes in the carbon dioxide (CO_2) content in the atmosphere (in parts per million, ppm) over the past 420 thousand years, according to the data obtained by drilling Antarctic ice. Four glacials and interglacials (with the names used in the Russian stratigraphy) are shown.

Events of the Quaternary Period and the Role of Carboniferous Limestone 53

The Likhvin interglacial, 420,000-395,000 years ago, which followed the Oka glaciation, was very extensive and characterized by a warmer climate than the modern climate and a large distribution of deciduous forests; in the southern half of the East European Plain, soils close to the soils of the subtropics dominated. In Western Siberia, this interglacial is called Tobolsk, in Northern Europe – Holstein, in North America – Yarmouth. Evidence of man's appearance on the East European Plain belongs to this period, but this evidence has not yet been confirmed for the Moscow region.

Starting from 300,000 years ago (with a peak about 250,000 years ago), a new powerful glaciation came – the Dnieper. The ice moved far to the south along the valleys of the Dnieper and Don. The thickness of the ice reached two or more kilometers. The ice spread from two centers: the main one, covering Scandinavia, and the additional one, which occupied the Polar Urals. In the European part of Russia, glaciers descended in the south by two giant invaginations – the Dnieper, moving along the Dnieper lowland to the latitude of the modern city of Dnipro, and the Don, moving along the Oka-Don plain to the mouth of the Medveditsa River, a tributary of the Don, starting in the Saratov region. The Central Russian Upland, which separated the two branches, was partially covered with ice only north of the city of Oryol. To the east, the southern border of glaciation went north along the western slopes of the Volga Upland and crossed the Urals.

After the Dnieper glaciation, the Odintsovo interglacial began, in which the climatic situation was close to the modern one. The Odintsovo interglacial was followed by the Moscow glaciation (170-130 thousand years ago). Its border ran from the southwest to the northeast of the Moscow region and divided it in half.

The Mikulin or Mikulino (known as Eemian in northern Europe) interglacial lasted from 130 to 115 thousand years ago. It is very interesting to compare it with the warm climate of the Holocene, i.e., the last 12 thousand years after the Younger Dryas. During the Mikulin interglacial, it was warmer than now. Oaks and hazels grew in northern Karelia, hippos and rhinos lived at the latitude of the Thames and the Rhine, and the North Pole was ice-free in summer. All this went on for 15,000 years, during the second half of which the temperature slowly dropped, and then began a sharper slope towards a new ice age (although it was still relatively warm until 70,000 years ago). If the glacial-interglacial cycles repeat the same way, we will live in a warm climate for about three thousand years until a new ice age begins. However, anthropogenic activity, which led to the release of huge amounts of greenhouse gases (carbon dioxide, methane, nitrous oxide), may prevent such a scenario.

7.2. Quaternary Development of the Oka Floodplain

The Kasimov territory was completely covered by ice during the Oka and Dnieper glaciations. The Moscow glaciation did not reach the Kasimov area, stopping to the north, at the latitude of Moscow, and only its melt waters reached the Kasimov area. Further north was the southern border of the Valdai glaciation. During the periods of cold snaps, tundra-steppes developed at the foot of glaciers and served as a food base for herbivores (mammoths, woolly rhinos, horses, saigas, deer, etc.). More than 20,000 years ago, ancient people came to the territory, armed with stone and bone tools, following behind herds of herbivores. Paleolithic sites of archaic humans on the Russian Plain are quite rare; one of them is the Sungir (at the tributary of the Klyazma, a tributary of the Oka, near Vladimir, a kilometer from Bogolyubovo), and the other is near the city of Spassk-Ryazansky. The age of the Sungir site is approximately 25,000 years (according to some sources, 30,000 years). There are many prehistoric sites in the Kasimov area, but they are all Neolithic (after the last ice age), and their age is no more than 4 thousand years.

Quaternary glaciations caused the formation of a terrace above the floodplain, on which a pine forest is located. It is composed of sandy and alluvial sand deposits, the total thickness of which ranges from 6 to 20 m. The pine forest on the left bank of the Oka was formed as a result of the last glaciations.

The glaciers left behind moraine loams with pebbles and boulders of various rocks. These include granites, gneisses, and quartzites of the Proterozoic basement brought from the Baltic Shield, Carboniferous dolomites and limestones, and chalk sandstones. The Dnieper glaciation left the most noticeable traces, as a result of which the thickness of the moraine reaches 15 meters with strongly prevalent sanded loams with lenses of sands. Sometimes, the thickness of the sediments reaches 100 m, and in watersheds, it usually does not exceed a few meters. Glacial boulders were found by my grandfather, Fyodor Dashkov, in the area of the Zybyonka stream, a tributary of the Tashenka River, above the village of Davydovo. He also reported on the discovery of mammoth bones in this place, as noted in the monograph "Archeology of the Ryazan Land" (Mongait, 1975) edited by Alexander

Mongait. Around this stream, there are sandy Quaternary deposits on which pine grows. We called this place Bugry (Hills) and collected mushrooms there.

In the book "Archeology of the Ryazan Land" (Mongait, 1975, pp. 259-260) (article by A.A. Mansurov and O.N. Bader "Archaeological map of the Kasimov environs"), the following data on the findings of Pleistocene animal bones are given:

> "Near the village of Baishevo in January 1930, mammoth bones were found, which were taken to the exposition of the Kasimov Museum.
>
> On the banks of the Zybyonka River between the villages of Zhdanovo and Davydovo, before the imperialist war, mammoth tusks were found, which were handed over to the police by the finder. The information was reported in 1923 by the Kasimov teacher F. Dashkov.
>
> Near the village of Popovka, a bison skull was found in the Cherny Rog stream in the 1910s.
>
> Near the village of Savino in January 1930, a mammoth sacrum was found.
>
> Near the village of Tashenka in 1929, while breaking stones in a quarry, workers discovered mammoth bones.
>
> In June 1926, near the village of Telebukino, during the extraction of stone, workers found four large mammoth bones and one tooth weighing 7.5 pounds (3 kg) at a considerable depth in a layer of white clay. In the area around Kasimov, the bones of a woolly rhinoceros were found."

After the last (Valdai) glaciation (11.7 thousand years ago), the Holocene epoch started, the climate stabilized, and human civilization emerged. Since the beginning of civilization roughly coincides with the start of the Holocene epoch, it has even been proposed to introduce a Holocene calendar, adding 10,000 years to our usual chronology. The Holocene calendar was introduced by Cesare Emiliani in 1993. This author also proposed adding the "human era" or Holocene era (HE) to the date of the Holocene calendar.

Most of the soils of Central Russia were formed in the Holocene. Holocene sediments include peatlands, fluvial alluvium, as well as many lake sediments. Peat bogs are often the result of an overgrowth of lakes, and sapropels (organic-mineral lake sediments) usually lie under the peat deposit. Holocene alluvial deposits include floodplain sediments, represented by sands, sandy loams, loams, and peat bogs formed on the site of oxbows.

At present, the Oka is an important physical and geographical boundary – the border between the forest and forest-steppe zones passes along its valley.

On the high right bank of the Oka, there is almost no forest; only in some places, the remains of oak forests and birch splits are preserved. There are many ravines formed by turbulent spring meltwater and flood streams. On the left side of the Oka valley, sandy terraces were formed. They are usually covered with pine forests, often with an admixture of linden and maple.

The left-bank bay of the Oka, which is an oxbow lake (Figure 9), connects with the river opposite the village of Vasilyovo, just opposite the ravine, in which I found many fossil brachiopods and other organisms. Here again, the Kasimov Age deposits appear (its Dorogomilovsky horizon), while upstream around Pervo and Baishevo to Tashenka come the limestones of the Myachkovsky horizon of the Moscovian Age. The banks of the Tashenka River in the lower reaches are composed of Myachkovsky limestone, while from the village of Davydovo, the upper reaches of the Tashenka are composed of Dorogomilovsky limestone of the Kasimovian Age.

The oxbow lake, on the one side bordered by a sandy terrace formed during the Valdai glaciation, and on the other by flood meadows on the alluvium of the Holocene epoch, is a beautiful place (Figure 9). I remember sailing on it on a boat with the teacher of physics and mathematics and my grandfather's friend, Alexander Petrovich Sotnikov. White water lilies (*Nymphaea alba*) and water caltrop (water chestnut) (*Trapa natans*) grow there in high abundance (Figure 9). Prehistoric people of the Stone Age and of later times relied significantly upon the fruits of water caltrops to supplement their diet, and, in times of crop harvest failure, water caltrops could be the main food component. In some areas, the plant played a role similar to that of the potato in our time. Archeological studies show that in pre-Mongol Russia, water caltrops were a widespread food. They were eaten like chestnuts, or dried, pounded, and added to flour. In China, water chestnut has been cultivated for at least three thousand years. The oxbow lake was once part of the main channel of the Oka, which flowed closer to the terrace of the pine forest. My grandfather heard from old-timers, and those from their grandfathers, how the Oka changed course, breaking through it in a new place. From the old channel, there is also a lake in the meadows a little higher upstream. There I found yellow irises and other coastal plants.

The last two hundred years has now been recommended to be defined as a separate epoch, the Anthropocene. It corresponds to industrial civilization when the human impact on nature is multiplied. In the area near Kasimov, an important role was played by developing limestone quarries, thus contributing to the appearance of the plant species originating from other regions. Many new plants, including those originating from North America, became

Events of the Quaternary Period and the Role of Carboniferous Limestone 57

widespread along the banks of rivers. In connection with human activity, we will concentrate on the epochs of the Holocene and Anthropocene a little later, and now we will characterize the unique vegetation of the banks of the Oka – the Oka flora, which interested me from childhood.

Figure 9. Photos taken at the oxbow lake of the Oka River by the author in 1975. Upper left – the view of the oxbow lake; upper right – pine forest on the sandy terrace near the oxbow lake; bottom left – *Nymphaea alba*; bottom right – *Trapa natans*.

Chapter 8

The Oka Flora on the Carboniferous Limestone

8.1. The Explorers of the Oka Flora

Ivan Alexeyevich Dvigubsky (1771-1840) (Figure 10), who was the rector of Moscow University in 1826-1833, noted the unique nature of the plants growing on the banks of the Oka River in 1828. He published the first overview of the flora of the Moscow region, which included 924 species. However, the real discoverer of the phenomenon of the Oka flora was Nikolai Nikolaevich Kaufman (1834-1870) (Figure 10), one of the first botanical geographers of the Russian Empire. He defined the Oka flora in 1861 and provided a detailed description. Nikolai Kaufman was an extraordinary person, as were his teacher Karl Frantzevich Roullier (1814-1856) and his student Pyotr Feliksovich Mayevsky (1851-1892). These people, despite their short lives, made a remarkable input in science. They were prominent representatives of the great scientific tradition formed in Russia in the 19th century. Nikolai Kaufman first taught natural history at the second Moscow gymnasium; in 1860, he started teaching at Moscow University, where he became a professor in 1863 and director of the university botanical garden. He also taught a course in plant physiology at the Petrovsky Agricultural Academy. His book "Moscow Flora" contains characteristics and keys to identify 941 species. It was the first original generalized floristic study in the R0ssian language. Kaufman was the first to raise the question of the natural causes for the entry of the southern steppe species into the forest belt, linking this with the role of the river in the settlement of plants. Kaufman bequeathed his herbarium to the university; his collections have survived until now. Since his student years, Kaufman suffered from a severe, undiagnosed illness, from which he died at the age of 36.

Figure 10. The florists who contributed to the study of the Oka flora. All images are in the public domain.

Kaufman's worldview was formed under the influence of Johann Wolfgang Goethe's natural science concept of metamorphosis. A peculiar monument to Kaufman became the Prioksko-Terrasny Nature Biosphere Reserve on the opposite bank of the Oka River from the science city of Pushchino, which I visited when he was a student. Kaufman studied the Oka flora in this place and discovered natural communities of steppe plants growing as enclaves far away to the north from the area of continuous distribution of the steppes. Since 1923, Associate Professor of the Moscow State University Pavel Alexandrovich Smirnov (1896-1980), who published the monograph "Flora of the Prioksko-Terrasny State Reserve" (1958), continued to study the Oka flora. The Oka flora of those places is a fragment of steppe meadows located in the areas of the high floodplain of the Oka, on the first terrace above the floodplain along the edges of dry steppe forests, and in some places under their canopy. The Oka flora also includes plants of southern origin growing on limestone outcrops. After Kaufman, different and sometimes opposing views were suggested on the origin of the Oka flora. Dmitry Litvinov (1902) put forward the idea of the relic glacial origin of the Oka flora, and Valery Ivanovich Taliev (1895) formulated the concept of its anthropogenic origin.

Before moving on to the phenomenon of the Oka flora, let us note the importance of Nikolai Kaufman's teacher, Karl Roullier, and his student, Pyotr Mayevsky. In my school times, I read about Karl Roullier in the wonderful book by Nikolai Plavilshchikov "Homunculus" and then the biographical book of Semen Mikulinsky. Karl Roullier developed a comparative-historical method for studying the organic world. He founded Russian evolutionary paleontology, was a remarkable zoologist and ecologist, and even engaged in what was later called bioethics. Roullier's work on the study of animal instincts and their mental activity laid the foundations of the evolutionary direction in zoopsychology.

Pyotr Mayevsky's guide "Flora of Central Russia" was my reference book in my school years when I spent summer holidays on the banks of the Oka. At first, I used Mark Neustadt's guide from my grandfather's library, another book that contributed to my interest in botany was the "Botanical Atlas" by Boris Shishkin, and then the associate professor of Voronezh University Galina Sergeevna Erdeli (1926-2020), offered me the supplemented ninth edition (1964) of Mayevsky's guide.

Pyotr Mayevsky edited the second edition of Kaufman's "Flora of the Moscow Province." Shortly before his death, he wrote in one of his letters: "It is terrible to bury your knowledge in yourself. I, at least, would like to chatter about my knowledge like a magpie. Well, isn't it painful to realize that I am the only one in the world who knows how beautifully, cleverly built [is] a watermelon?" Pyotr Mayevsky was the home teacher of the famous book publishers, brothers Mikhail and Sergey Sabashnikov. One of them, Mikhail, recalled: "Mayevsky seemed to be a man destined to compile a Russian botanical guide. A deep connoisseur of Russian flora, he had just finished editing a posthumous edition of Kaufman's Flora of the Moscow Province. The usual and, perhaps, the only path open to a Russian scientist at that time – the path of the professorial chair – was closed to Mayevsky. Pyotr Feliksovich was a hunchback. With constant interruptions of the heart, suffering from shortness of breath, quickly tired when moving or excited, he was unable not only to read the usual two-hour lecture in front of an audience but also to hold on to the people for a long time. The deskwork of a scientist or a writer, which could be done at home in complete tranquility – that is what he was able to do. We invited him to compile a large guide to plants..."

Regarding the Oka flora, we should pay tribute to two Russian botanists, Dmitry Litvinov and Valery Taliev (Figure 10). Their views on the origin of the southern flora along the banks of rivers were exactly the opposite, but the truth is paradoxical and, obviously, includes even mutually exclusive explanations. Dmitry Ivanovich Litvinov (1854-1929), together with Vasily Yakovlevich Zinger (1836-1907), was the discoverer in 1882 of Galichya Gora – a unique place for the growth of relict flora on the steep bank of the Don River, formed by Upper Devonian limestones. Dmitry Litvinov put forward a hypothesis about the relict origin of pine forests on the chalk deposits in the European part of Russia. According to Litvinov, with the retreat of glaciers, part of the flora on their outskirts began to spread to the north; some died out on the plains of European Russia, preserved in neighboring hilly areas, and some have survived to our time in the forests on river terraces. Litvinov also denied the independent nature of the Arctic flora, considering it a derivative of the alpine flora of the southern areas. Valery Ivanovich Taliev (1872-1932) adhered to the opposite point of view. He considered the anthropogenic factor to be decisive in the history of the spread of vegetation. The peculiarity of the flora of the Crimea, the peculiarities of the chalk outcrops, the forestlessness of the steppes, and the disappearance of rare plants, he associated with the activities of man and partly animals.

It is necessary to say a few words about Vasily Zinger (Figure 10) in connection with the role of Nikolai Kaufman as his mentor in botany, and his son Alexander, who was a physicist, interested also in botany (he was the author of the popular book "Entertaining Botany," which was in my grandfather's library) and a friend of Leo Tolstoy. Alexander Zinger quoted his father's words: "When I watched how Kaufman collected and explored plants, when I listened to his stories, my eyes were literally opened: the grass, the forest, and the soil appeared to me in a completely new light, full of the deepest interest." Several plants are named in honor of Vasily Zinger, including the cereal *Zingeria*. Some species of this genus have the smallest number of chromosomes in higher plants (the diploid set – four chromosomes). Vasily Zinger was one of the brightest representatives of the Moscow philosophical and mathematical school. He developed the ideas of a systematic approach, based on the principles of the philosophy of Immanuel Kant.

The author's first research at school age was devoted to the Oka vegetation, and at that time, the author accepted Taliev's ideas about the anthropogenic factor in the settlement of plants along the banks of rivers. Later, while studying at Voronezh University and having summer practice in the Galichya Gora Nature Reserve, as well as being in Pushchino on the banks of the Oka River (near the Prioksko-Terrasny Reserve), the author realized that Taliev's views were sometimes too radical, as it was difficult to deny the relic nature of some representatives of the Oka flora. Nevertheless, the caliephilicity of limestone plants was the main factor contributing to their dispersal, and their relict origin since the last glaciations is often not obvious. In addition, the distribution of plants originating from North America along the banks of rivers, as well as the recent appearance of new species of southern origin at the sites of limestone development, partly confirm Taliev's views. Later, I learned that the limestone outcrops on the banks of the Tashenka River (a tributary of the Oka), which attracted my attention in connection with the growth of the Oka flora, are associated with limestone excavation in the 19th century (this is described in Dolotov's article of 1999) (Figure 11). These places of vegetation of the Oka flora are obviously of anthropogenic origin.

Figure 11. The valley of the Tashenka River (left) and the limestone slope (right) on which the representatives of the Oka flora grow. Author's photos (2013, 1975).

My grandfather, Fyodor Andreevich Dashkov (1900-1983), attended a lecture of Taliev, probably when Taliev visited the Ryazan Pedagogical Institute in the late 1920s. He noted the extraordinary personality of Professor Taliev and regretted his early death from cancer (Taliev passed away one day before his 60th birthday). My grandfather explored the settlements of the early Iron Age and participated in the creation of the local Kasimov history museum. This contributed to my interest in the relationship between the spread of the Oka flora and human activity.

Let's say a few words about the life of Valery Taliev. He was born on February 22 (February 10 of the old style) 1872 in the small town of Lukoyanov, Nizhny Novgorod province, in a Mordovian family. His father was a teacher. In 1883, he graduated from the Sergachev District School and then studied at the gymnasium at the Nizhyn Historical and Philological Institute. After receiving a graduate certificate in 1890, he entered the natural history department of the Faculty of Physics and Mathematics of Kazan University, from which he graduated in 1894 with a diploma of the first degree. After that, he immediately entered the 3rd year of the Medical Faculty of Kharkiv University, and in 1897, he received the title of doctor with honors. Until 1899, he served as a military doctor. In the same year, he passed the master's exams at Kharkiv University, and in 1900, he became a privatdocent of Kharkiv University. In 1916, he defended his doctoral dissertation at Petrograd University on the topic "Experience in studying the process of speciation in wildlife." In 1917, he was appointed as a dean of the natural history department of Kharkiv People's University. At the end of 1918, Taliev was invited to Moscow as an elected professor of the Petrovsky (later Timiryazev) Agricultural Academy.

The anthropogenic factor in the settlement of plants is devoted to Taliev's Master's thesis "The Flora of the Crimea and the Role of Man in its Development" (1902). In subsequent works, Taliev developed the ideas initially formulated in his dissertation. Taliev believed that man actively transformed spatial-temporal connections in the biosphere. This is manifested in plant dispersal and the transformation of relationships between different vegetation types. Spatial and natural barriers in the settlement of plants lose their importance with the expansion of man. New plants usually come from the southern areas, which indicates an active process of speciation in southern latitudes. Later, Sergei Meyen (1987) investigated the process of progressive evolution in tropical and subtropical regions and the subsequent spreading of plants to northern latitudes. He defined this process as a phytospreading (Sharov and Igamberdiev, 2014).

Taliev denied the relict nature of plants of limestone and chalk deposits. He pointed out that the sand and chalk outcrops were formed by human and animal activities that destroyed the upper layers of the soil. Pine forests were retreating, exterminated by man, and their remnants were preserved on chalk and limestone outcrops. In the proximity of pine forests and on hard and dry chalk deposits, plants can withstand drought conditions. These are often mountain plants, natives of the Crimean, Caucasus, and Altai mountains, brought by man to chalk outcrops, and, according to Taliev, not relics at all.

8.2. The Postglacial Spreading of Plants

In the Holocene, man definitely facilitated the postglacial distribution of plants. This distribution took place both from the southern areas and from the habitats of refugia flora, e.g., in the places bypassed by a glacier. These patterns are confirmed in the latest studies in the study of plant pollen, which has been preserved for thousands of years (Alsos et al., 2005). Thus, the ideas about the relict nature of the limestone and chalk plants do not contradict the primary role that man played in their settlement. It is now evident that the cold and dry climate during the Quaternary period contributed to the formation of calciphile soils. The distribution of calciphile species of the Central European flora is the result of this recent process, which has been most intense in the elevated places of the periglacial zones. At the same time, the species that lived on acidic soils and did not have calcium tolerance became extinct faster than calciphiles (calcicoles) in the cold climate of the ice ages.

The available data show that the effect of the first human settlements along the banks of rivers was substantial. Agricultural activities along the banks of rivers led to the depletion of soils. Indigenous deciduous forests along the banks of rivers were destroyed and replaced by birch and aspen, which can be considered as weed tree plants. Deforestation and agricultural cultivation in the late Holocene caused increased spring flooding, erosion of steep banks, and alluviation of flood meadows. The first human settlements significantly changed the soil profile. These changes can be traced in central Russia to 4-5 thousand years before the present.

In most cases, the hydromorphic soils and yellow soils of the Middle Holocene were replaced by the gray forest soils of the Late Holocene. Then, an important factor was the freeing of areas along the banks of the rivers from the forests that grew there. Deforestation and agriculture in the river basin in the last 700-900 years have caused intensive erosion of the slopes and alluviation (washing of sediments) of river valleys. As a result, gray forest soils were buried under recent alluvium, on top of which a new alluvial soil layer was formed. Buried soils can be found on the sites of human settlements. They correspond to different settlement layers, from the Neolithic to the late medieval period.

In 1980, Taliev's hypothesis received an unusual and even more radical development in the article by Kostenchuk and Tyuryukanov (1980). The authors established a direct connection between the emergence of the Oka flora and the war campaigns of nomads in Russia. In their opinion, the path of the Batu Khan troops in the winter of 1237-1238 completely coincided with the resettlement of steppe plants along the Oka. Batu Khan's army consisted of 120,000 to 140,000 men and at least 200,000 horses. Harvesting food for so many animals took place near the current town of Voronezh on the Don River. In winter, nomads moved along the banks of the Oka, Moskva, and Klyazma rivers, sowing southern forms in the floodplains of rivers.

Obviously, both the anthropogenic and natural factors played a role in the settlement of plants and their subsequent evolution. Taliev always emphasized the multifactorial nature of the evolutionary process. He was the first to draw attention to the importance of the anthropogenic factor and substantiate it with specific examples (Igamberdiev, 2021). Steppe plants possess a long overall length of their roots, which allows them to compete successfully with trees. On the other hand, trees can overcome grass resistance by providing and benefiting from the heterogeneity of soil resources. The role of man may be to shift the equilibrium between steppes and forests, and even relatively small destruction of forests can lead to the spread of steppes and steppe vegetation over wide areas. "Exposing the vast expanses once covered with dense forests, man was a powerful companion of the steppe in its age-old struggle against the forest," Taliev (1895) wrote. "The chalk outcrops with all their attributes (steppe flora) can be and are, in fact, the subsequent result of human influence." In some cases, the studies of ancient settlements (for example, in the Urals) directly confirm the fact that they played a direct role in replacing forests with steppe vegetation. The connection between ancient settlements and changes in soil cover was analyzed in detail in the works of Boris Pavlovich Akhtyrtsev (1929-2008), a professor at Voronezh University, with whom I had a chance to communicate. He pointed to the important role of the anthropogenic factor in changing soil cover and vegetation types.

Taliev considered the weeds of central Russia to be newcomers, having a southern origin and formed in a drier continental climate. Hence, the characteristic feature of the structure of weeds is adaptability to moisture retention, to reduce evaporation. At present, it is obvious that many weeds are indeed of southern origin. Many of them have a C_4 type of photosynthetic metabolism and their closest relatives live in the southern zones. In addition, many weeds in origin are confined to the centers of origin of cultivated plants, from where they spread to the north with the development of agriculture.

Taliev identified three types of weeds, which are differently adapted to new habitats. The first type (e.g., thistles) inhabited cliffs, wastelands, and garbage pits, the second type (e.g., dandelion or plantain) has adapted to living in yards, pastures, and fallow fields, and the third type (e.g., cornflower) has a growing season that coincides with the development of cultivated plants.

"There is a profound analogy between the vegetation of rocky outcrops and weeds ... In fact, what are weeds? The easiest way would be to answer that weeds are the plants brought in by man that find suitable soil conditions in the vicinity of human settlements... But this answer will indicate only one feature, an insignificant one. In reality, the way of appearance distinguishes weeds into a special category. Being an alien element in our flora, weeds settle only where the barriers to the penetration of plants are reduced. Such places are those points in which man, in one way or another, exterminated or suppressed native vegetation. Chalk vegetation also belongs here" (Taliev, 1896).

I wrote about specific representatives of the Oka flora in the area of the village of Pervo and the vicinity of the Tashenka River in two articles during school time. I am presenting one of them in the appendix of this book.

Chapter 9

Holocene: First Human Settlements and Neolithic Agriculture

The first settlements on the banks of the Oka near Kasimov belong to the Neolithic period and are associated with the settlement of Finno-Ugric tribes during the Holocene. The Holocene epoch is divided into three ages – Greenlandian (approximately 11.7-8.2 thousand years ago), Northgrippian (8.2-4.2 thousand years ago), and Meghalayan (4.2 thousand years until now or the beginning of the Anthropocene). The exact timings of the Holocene ages in the Holocene calendar (Holocene era, HE) are: the Greenlandian – 300-3854 HE, the Northgrippian – 3854-7750 HE, and the Meghalayan – from 7750 HE). The Neolithic settlements around Kasimov belong to the end of the Northgrippian and the Meghalayan ages.

The earlier Neolithic sites were located on the lower terrace of the left bank of the Oka. Close to Kasimov is the Neolithic site near the village of Popovka, the age of 4-2 thousand years BCE. The ceramics in this site had a pit-comb pattern. Late Neolithic ceramics (2 thousand years BCE) had the predominance of the comb pattern with a less dense filling of the ornamental field. At the end of the second millennium BCE, a type of ceramics appeared on the Oka with textile prints on the outer surface, made with flint tools.

At the turn of the third and second millennium BCE, the arrival of the tribes of the Volosovo culture on the Oka took place. According to Otto Bader, the ancestors of the Volosovites migrated from the Ural-Kama region. Around the beginning of the second millennium BCE, the Volosovo culture was replaced by the Fatyanovo and the related Balanovskaya culture (although representatives of these cultures may not have spread around Kasimov), and then (from the middle of the second millennium BCE) by the Pozdnyakovskaya culture. The appearance of the Pozdnyakovskaya settlements on the Oka is associated with the northward movement of the tribes of the Srubnaya culture from the forests and steppes of the Upper Don and their assimilation of the local population. This movement was facilitated by the shift to the north of the landscape zones that took place in the Meghalayan age, the onset of the steppe and forest-steppe on the forest, and the corresponding movement of the Oka tribal forest hunters and anglers who retreated along with the forest. Thus, the climate changes in the Holocene were the most important factor determining the migration of peoples.

The settlements on the terrace of the left bank of the Oka became possible in conditions of a drier climate than the modern one when the floods of the Oka did not cover the floodplain and did not reach the level of the settlement. These conditions were particularly consistent with the xerothermic time at the beginning of the Meghalayan age. It corresponds to the second millennium BCE. After that time, the climate was marked by significant humidification and, consequently, a change in the Oka regime. The lower Neolithic sites, affected by the Oka floods, were abandoned by people and were gradually covered with thick alluvial deposits. This happened approximately in 800 BCE.

Closer to the middle of the first millennium BCE (the 7th century BCE), a new culture of pastoralists and forest farmers was formed around Kasimov. It was the Gorodetskaya culture, which was related to the Dyakovskaya culture. The representatives of this culture were engaged in cattle breeding and agriculture, they had a knowledge of how to make harpoons and arrowheads from iron. The Gorodetskaya tribes belonged to the ancient Finno-Ugric population of the central part of the East European Plain. They moved their settlements or hillforts (called "*gorodishche*" in Russian) from floodplains and lowlands to high banks. These settlements belong to the Dyakovskaya culture. People settled on steep banks, and the traces of their life were preserved primarily in these ancient settlements. There are several settlements in the described area. One of the best preserved is the Baishevsky, near the Tatar village of Baishevo, opposite which, through a ravine, closer to Pervo, is the Russian village of Shemyakino. The hillfort has a well-preserved steep rampart, later there was a Tatar cemetery with the monuments tilted to the east, in the direction of sunrise. From the high Baishevsky hillfort, on which the settlement is located, there is a beautiful view of the Oka, flood meadows, and Kasimov (Figure 12) similar to the one described in the story "Kuzma Roshchin" by Mikhail Zagoskin.

The famous Kasimovian historian and architect Ivan Gagin described the Baishevsky settlement in the 1830s (Figure 13). Otto Bader examined it in 1932 and noted the destructive work of the Oka, which washed away a part of the coast between the Baishevsky settlement and the river and significantly eroded the base of the cape on which the settlement is located. The development of a limestone quarry also contributed to the destruction. The entire site and part of the rampart were, at that time, under the densely located graves of the old Tatar cemetery overgrown with old birch trees, but now only a few Tatar monuments still exist there.

Figure 12. The photos of Oka, the ravine, and the limestone quarry at the Baishevsky Neolithic settlement. Author's photos (2009, 2013).

Another settlement (the Zhdanovsky hillfort) is located on the left bank of the Tashenka River. According to Alexei Mansurov and Otto Bader (see Mongait, 1975, p. 283), "the Zhdanovsky settlement was described back in 1874. In 1925, the settlement was surveyed by F.A. Dashkov, who dug a pit in it and found out that the settlement belonged to the early Iron Age. The thickness of the cultural layer is 70 cm."

The culture of the Dyakovskaya settlements lasted on the territory of Kasimov for a period of more than a millennium, from the second half of the first millennium BCE to the first half of the first millennium AD. In the book "Archeology of the Ryazan Land" (Mongait, 1975, p. 315), we read: "At that time, cattle breeding became predominant. Hunting and fishing fade into the background, although they continue to play a certain role in economic life. Hoe-slash farming was developing, but it was not predominant. The patriarchal-tribal way of life was still quite strong. Obviously, the inhabitants of one settlement constituted the main production unit – the patriarchal family.

The settlement was not only a military fortification but also a place of constant protection of herds."

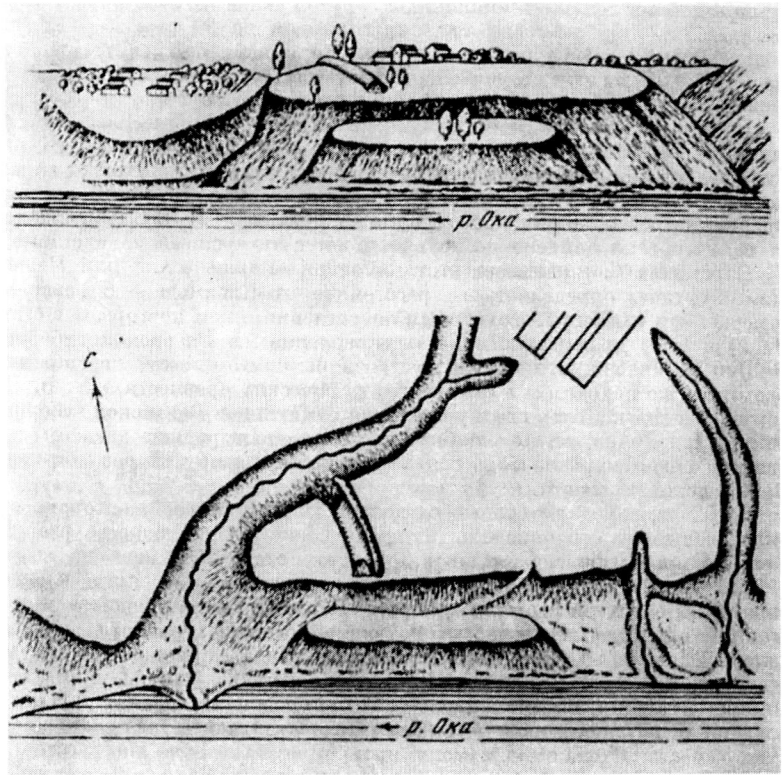

Figure 13. Baishevsky Neolithic settlement drawn by Ivan Gagin in the early 19th century. Public domain.

In the village of Pervo, at the exit to the Oka on a steep bank (a place called "Paltso"), there was also a settlement of a smaller size than the Baishevsky hillfort. On the high places between the ramifications of the Guzhovka ravine (around the village of Pervo), traces of ancient anthropogenic activity are also noticeable, as pointed out by Fyodor Dashkov. He also assumed that the unusually flat surface of the hill Sechka (the name means "truncated") between the ravines, very close to his house, indicates that people of the Neolithic times lived there. Maybe a sanctuary of the pagan god Perun, from which, according to some assumptions, the name of the village of Pervo originated, was there.

On the slopes of the Baishevsky hillfort, intermediate wheatgrass *Thinopyrum intermedium* (Host) Barkworth & D.R. Dewey, previously classified as *Elytrigia intermedia* (Host) Nevski, grows in abundance. Its spikes are seen in the photo (Figure 14). I found it in 1974 and suggested that Neolithic people could use it as an agricultural plant. It is known that this species was cultivated as an ancient grain crop in Western Asia (Morrison, 2004). This plant is similar to ordinary wheatgrass, *Elytrigia repens,* which is now classified as *Elymus repens* (L.) Gould, but less common and is grown mostly in southern areas. Wheatgrass is a weed related to cultivated cereal plants. The presence of intermediate wheatgrass near the Baishevsky settlement should be connected with Neolithic farming. This is possible because it is a promising perennial cereal crop and one of the most productive forage species. Like bread wheat, intermediate wheatgrass is allohexaploid ($2n = 6x = 42$), possessing many desirable agronomic traits that make it an invaluable source of genetic material useful in wheat improvement (Mahelka et al., 2011). It is nutritionally similar to wheat, and the grain can be ground into flour and used for food products. Now it is considered the most promising species based on flavor, ease of threshing, large seed size, resistance to shattering, lodging resistance, ease of harvest, and perennial growth (Wagoner and Schauer, 1990). Currently, some products from this plant are marketed under the trade name Kernza (Becker, 1991).

It is known that several plants that were weeds in the primary agricultural centers, upon spreading to higher latitudes, became important agronomic cultures. These plants include rye and oats, which were the weeds of wheat crops in the Near East. During spreading to the northern regions, they became cultivated as crops (Schreiber et al., 2018). Intermediate wheatgrass can also be considered a weed plant of southern origin that became cultivated in Neolithic European settlements at higher latitudes. Its growth at the place of Baishevsky settlement may indicate this possibility.

In the second half of the first millennium AD, the Finno-Ugric Meshchera tribes, related to the Mordovian nationality, appeared on the territory of the Kasimov region. Around the tenth century, the intensive penetration of Slavic tribes into this region took place. The Slavic tribe Vyatichi actively developed these places, and in the 11th and 12th centuries, the Slavic people who were closer to Krivichi than to Vyatichi came from the Vladimir and Suzdal lands. The Meshchera tribes absorbed the elements of the Slavic culture. The development of the Meshchera lands by the Slavic tribes took place rather slowly and took several centuries, until the invasion of Batu Khan. The main occupation of the Slavs was agriculture. As Alexei Mansurov and Otto Bader mentioned, "this process proceeded slowly. While to the south and west, the features of the future Russian nationality were intensively formed, here, in the wilderness of the Meshchera forests, there were still native tribes. Perhaps this isolation from the Moscow centers explains to a certain extent the fact that the Tatar Khanate was established here. A semi-wild country, with a significant percentage of the pagan population, was easier to give in to the possession of the Muslim ruler than the areas with an indigenous Russian Christian population" (Mongait, 1975, p. 315).

Chapter 10

Anthropocene and Recent Carboniferous Limestone Developments

The last 200 years are widely considered a new geological epoch – the Anthropocene. Paul Crutzen (1933-2021), a Nobel Prize winner for his study of the atmospheric ozone layer, proposed the term (Crutzen, 2002); however, the International Commission on Stratigraphy (ICS) and the International Union of Geological Sciences did not approve this suggestion. Vladimir Vernadsky (1863-1945) was more radical, he believed that with the advent of human civilization, a new geological era began – the Anthropozoic (Vernadsky, 1926). The Anthropocene began with the Industrial Revolution (the development of the steam engine by James Watt in 1784), although there is a suggestion of counting it from the mid-twentieth century when the tests of nuclear bombs represented its marked geological signature. It is characterized by a significant increase in greenhouse gases in the atmosphere due to human activity, as well as many other aspects of anthropogenic impact, such as industrial nitrogen fixation for fertilizer production, which already exceeds the pre-industrial level of nitrogen fixation.

The anthropogenic impact on the biosphere intensified in the 20th century, especially in its second half. Carbon dioxide increased from 280 parts per million (ppm) in the nineteenth century (pre-industrial level) to 425 ppm by 2025, and methane increased from 0.7 to 1.9 ppm. The lowest glacial concentrations of CO_2 were around 180 ppm and methane 0.4 ppm. The temperature has also increased by 1.5 degrees from the preindustrial times. The average global temperature in the last ice age (21,000 years ago) was 5-6°C colder than the preindustrial, but in the proximity of ice sheets, it was 10-15°C colder. The current fluctuation also coincides with entering a new (Modern) warm period (the previous ones – Medieval, Roman, Minoan – came at intervals of about a thousand years) (Figure 15).

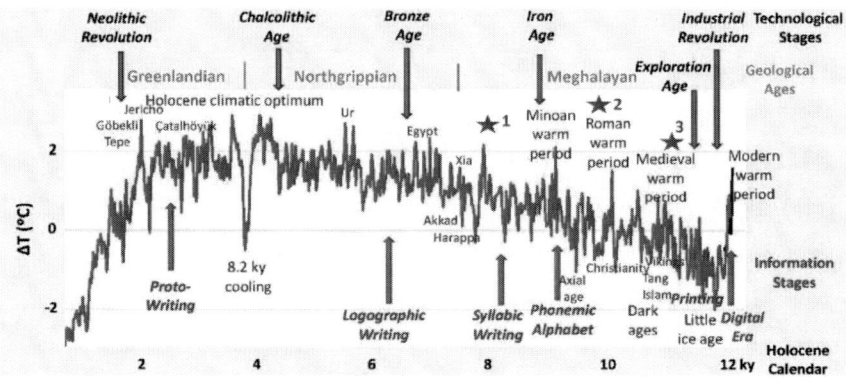

Figure 15. Changes in temperature (ΔT) over the past 10 thousand years (Holocene epoch and its three geological ages). The temperature curve of the Holocene in the Northern Hemisphere is based on the proxy data from Greenland ice cores (https://www.carbonbrief.org/factcheck-what-greenland-ice-cores-say-about-past-and-present-climate-change). The observational data for the 20th century (http://berkeleyearth.org/archive/data/) are presented in black. The main events in the development of civilization during the Holocene are shown, with the indication of technological and information stages. Asterisks mark the time of the foundation of the Oka floodplain Neolithic sites (1), the emergence of the Finno-Ugric culture (2), the spread of Slavic tribes, and the founding of Gorodets Meshchersky (Kasimov) in 1152 (3). The picture is drawn by the author.

My childhood was at the end of a cooling trend, characterized by a downward trend in temperature (between the 1940s and 1970s). I remember that in the summer holidays, the temperature was often around 13°C with rain almost every day. The hot summer of 1972, when forests were burning all around, marked the end of this trend. In the scientific and popular literature (I remember one in the magazine "Around the World"), the concept was often substantiated that human activity leads to cooling due to the release of aerosols. Later, this statement was marginalized, and the theory of global warming became prevalent.

Limestone has been mined around the village of Pervo for many years. Especially active were the developments in the 19th century, during the Industrial Revolution in Russia. As a child, I remember deep sinkholes in the limestone slopes caused by limestone mining, which were dangerous – it was possible to fall into them. The peasants, who attended seasonal work in limestone quarries, earned enough money to build solid stone houses, which are still preserved in the village of Pervo.

The detailed data on old limestone quarries in this region can be found in the article by Dolotov (1999). The author presents the data on the development of limestone in the Kasimov region downstream of the Oka below Kasimov. Following the information provided in this article, we will start with the current Kasimov limestone quarry and go down the Oka, and then go up the Tashenka River. The author refers to a local resident of the village of Volkovo, which is next to the current Kasimov quarry, above the confluence of the Tashenka and the Oka, who stated that on the site of the Kasimov quarry, there were previously small open-pit mines developed by the industrial cooperatives, and prisoners and captured Germans also worked there. The work was carried out without any mechanization, picks, or wheelbarrows. Developments of this type were also below the confluence of the Tashenka in a continuous strip between its mouth and the village of Baishevo. The author of the article examined the area of the village of Baishevo and described what was left of the old quarries that were developed without any underground work. But below the Baishevsky settlement, starting from the village of Pervo and to the village of Maltsevo, the traces of underground quarries stretch for more than 2 km along the right bank. This development was carried out from the 19th century to the 20th century, reaching its maximum scale in the late 1920s. On the high slopes of the shores, one can see traces of the lines of grinding ditches, dumps under them, dumping blocks, and landslides looming on top of all this. Ravines that cut through the shore to or below the level of the grinder line make it possible to divide the development field into separate blocks. The upper boundary of the development above the village of Pervo is covered with forest; however, there are sinkholes and ditches above the village of Pervo, up to the village of Shemyakino and the Baishevsky settlement. The traces of a small quarry were found further down the Oka, 750 meters below the village of Balushevo-Pochinki.

On the left bank of the river Tashenka, half a kilometer above the village of Tashenka, located on the banks of the Oka, dumps and eroded grinding ditches are also found, which, according to the author, indicates the ancient development of limestone. Further, a couple of kilometers up, near the village of Savino, large limestone developments are observed stretching for two kilometers with huge dumps under large grinding ditches, often eroded, indicating several mining sites in this area. On the right bank of the Tashenka River, near the village of Davydovo, dumps with grinding ditches above them, including overhead failures, were discovered. The valley of the Tashenka River here is a narrow incision with steep, high (40-50 m) slopes, completely forested and not visible from the side. All quarry systems inspected by Dolotov are located in the middle of the slope at an altitude of 20-25 m from the lowest elevation and have powerful dumps descending to the foot of the slope.

It was here that I found the largest number of plants of the Oka flora, including *Salvia pratensis*, *Carduus thoermeri*, *Geranium sanguineum,* and *Pyrethrum corymbosum*. I describe these plants in my article at the end of the book. It turns out that these limestone outcrops, which are abandoned quarries, are less than 200 years old, according to Dolotov, and the diversity of the southern steppe flora here may be quite recent. This indicates that Taliev was right in understanding the role of man in the spread of plants over the banks of rivers. In the areas of new developments of the Kasimov quarry, completely new plants appeared very recently, such as *Chaenorhinum viscidum*. I remember how, in the lake next to the quarry, I found an abundance of flowering *Elodea canadensis*. The flowering of this plant, having North American origin, is quite rare.

It is noted that the development of limestone was carried out near the village of Chinur, which has a Finno-Ugric name and is located on the banks of the Chinur stream, a tributary of the Tashenka. As Dolotov notes, opposite the lower end of the village on the right bank of the stream, there are traces of underground mining – short grinding ditches with overhead sinkholes. Directly below the mining site, the stream goes underground, flowing like a waterfall into a karst lake measuring 10 by 20 m, with steep limestone banks. Further down, the channel to the confluence of the Tashenka is a dry valley. The water of the stream never appears on the surface, dispersing in the groundwater horizon.

Thus, the relatively recent development of limestone along the Tashenka and Oka rivers has led to a change in the composition of the flora in this area and to the spread of calciphile plants. Thus, floristic changes occur quite quickly, and the spread of more southern plants under the influence of human activity is a fairly obvious fact.

In the town of Kasimov, there are many invasive weeds, the growth of which is also associated with the calcareousness of the soil. I collected several plants for the herbarium there, although I studied these plants in less detail than those that grow on the banks of the Oka and Tashenka near Pervo and nearby villages.

Chapter 11

The Town of Kasimov

The limestone deposits of the Oka played a major role in the history and architecture of the town, which was founded as Gorodets Meshchersky by the Suzdal prince Yuri Dolgoruky in 1152. The old town was situated on a cape formed at the confluence of the Babenka River and the Oka River more than a kilometer downstream from where the city center is now located, opposite the Stary Posad district. It became a stronghold of the settled Finnish tribes and Slavs and served as a fortress city to protect the borders of the Vladimir-Suzdal principality.

According to Ivan Gagin, the Mongols destroyed Gorodets Meshchersky in the 1370s. The town was restored later, not in the old place but higher up the river, where the city center is now located. In the New Encyclopedic Dictionary of Brockhaus and Efron (volume 21), it is noted that in the 14th century, there were the princes Meshchersky from newcomers who converted to Orthodoxy. The Mongols destroyed Gorodets Meshchersky in 1376. Then it was restored and ceded to Moscow. From 1376 to 1471, the city was called Novy Nizovoi.

A special period in the history of the city is from the middle of the 15th century to the end of the 17th century. In 1452, the Moscow prince, Vasily II the Dark, gave the city to Qasim Khan, who fled from Kazan. Qasim Khan faithfully served the Russian people, and for this, in 1471, the city was renamed in his honor. In connection with the formation of the Kasimov Tatar Khanate in the Meshchersky Territory, the capital of which was the city of Kasimov, Tatars from Crimea, Kazan, and other places arrived here. The Kasimov Kingdom, which was directly dependent on Moscow, lasted for more than 200 years. During this period, the Russian and Tatar cultures were intertwined. The appearance of the city and its environs changed; a minaret tower (1467), a mosque, and the mausoleums of Shah-Ali Khan (1555) and Afghan Muhammad Sultan (1649) were built there (Figure 16).

Figure 16. Architectural monuments of Muslim culture in Kasimov. Upper images: Khan's mosque (18-19th century) with minaret (15-16th century). Bottom images: Shah-Ali's (1556) and Afghan Muhammad's (1649) mausoleums. Author's photos (2009).

The German scientist and diplomat Herberstein, who visited Russia in 1526, in his work "Notes on Muscovite Affairs," briefly described his contemporary Kasimov, who was at that time under the control of Dzhan-Ali. The next mention of Kasimov belongs to Adam Olearius, who traveled with the Holstein ambassadors to Persia and visited Kasimov in 1636. In 1552, Kazan was annexed to Russia, and the local "tsars" moved into the category of landowners. In 1681, the Kasimov Kingdom ceased to exist even formally. Stone Orthodox churches began to be built in Kasimov. In 1700, one of the first to be built was the Epiphany Church (also called Yegoryevskaya) (Figure 17), and a year later, St. Nicholas Church. Ivan Alexandrovich Balakirev (1699-1763), who was a jester of Peter the Great, was buried near the Yegoryevskaya Church. He came from an old noble family, and his surname probably originates from the Tatar words "bala kire" (stubborn child). In 1722, on his way along the Oka River to the Persian campaign, Peter the Great visited Kasimov for the second time, and in his retinue was Ivan Balakirev. Having learned that the title of the ruler of the city was not occupied, Balakirev asked the tsar for permission to be called Khan of Kasimov. The tsar jokingly agreed, so the "khan" appeared again in Kasimov. Initially, this title was formal, but after Peter the Great's death, by decree of Catherine I, Balakirev received the right to own the former estates of the Kasimov tsars, the rank of lieutenant of the Life Guards, and the title of "Tsar of Kasimov." Ivan Balakirev's grave is located behind the altar of the Yegoryevskaya (Epiphany) Church (Figure 17).

Figure 17. Architectural monuments of Kasimov. Upper left photo – the panorama of Kasimov from the Oka River with St. Nicholas Church (1705), Alyanchikov's house (1810), Assumption Church (1775), and Cathedral of the Ascension (1864). Upper right – Annunciation church (1740). Bottom left – shopping arcades (1824), bottom right – Yegoryevskaya (Epiphany) Church (1700). Author's photos (2009).

At the end of the 18th century, the city began to transform, and a plan for its development was established. After the fire of 1828, when most of Kasimov burned down, stone buildings began to be built, many of them today are architectural monuments. A huge role in creating a new image of the city, as well as in preserving historical documents about it and its environs, belongs to Ivan Sergeevich Gagin (1767-1844). He was an architect and one of the first Russian local historians (Figure 18). Ivan Gagin designed many of the buildings that define the appearance of Kasimov, including shopping arcades and the central square. The descriptions, drawings, plans, and maps made by him are of enduring importance for the study of Kasimov's history.

Figure 18. Ivan Gagin (1767-1844). The portrait was made by V.F. Timm in 1862. Public domain.

Ivan Gagin was a Kasimov tradesman who lost his fortune in his youth and devoted his whole life to the study and development of the local region. Thanks to Gagin, we have accurate information about the location of the royal palace in Kasimov, a plan for the location of the fortress walls of the Kasimov Kremlin and Gorodets Meshchersky, and good drawings and plans of settlements, including the Baishevsky.

In 1854, the Ascension Cathedral (Figure 17) was erected in the city center according to the project of the provincial architect Nikolai Ilyich Voronikhin. At the end of the eighteenth century and the beginning of the nineteenth century, Cathedral Square was built. Nikolai Ilyich Voronikhin (1812-1877), a Ryazan provincial architect, was the nephew of Andrei Nikiforovich Voronikhin (1759-1814), the architect of the St. Petersburg Kazan Cathedral.

On the central market square, there is a beautiful Annunciation Church (Figure 17), and next to it is the house in which the doctor, Fyodor Alexandrovich Sokolov, the son of a priest, lived. My grandfather told me that in the 1920s, he conducted a skin graft experiment on himself, from which he died.

A tireless researcher of the history of the Kasimov region and the main creator of the local history museum of the city was Alexei Alexeevich Mansurov (1900-1941). My grandfather remembered him as a friend and regretted his death at the beginning of the war. Mansurov (Figure 19), who had severe myopia, joined the ranks of the people's militia, whose goal was to stop German troops on the outskirts of Moscow. Untrained, practically unarmed people were sent to inevitable death. The article by Bader et al. (1975) presents an overview of Alexei Mansurov's contributions and his biography.

Alexei Mansurov
(1900-1941)

Figure 19. Alexei Mansurov. The image is in the public domain.

Afterword

Now we end our "brief history of time" that we observed from Kasimov's frame of reference. To complete it, we add the text "The History of the Village of Pervo" in the Appendix, written by my grandfather Fyodor Dashkov. The history of one small village is thus placed in the context of the universal history of the Earth and humankind.

If we turn to global processes, then the Earth's geological history can be predicted more or less definitively for the next few hundred million years based on the tectonics of lithospheric plates. The East European Platform will continue to be at the center of the greater Eurasian continent, which Africa will join in a few tens of millions of years, forming the Mediterranean Mountains on the site of the current Mediterranean Sea. Further movement of the continents will lead to the formation of a new supercontinent, Pangea Proxima, in two to three hundred million years.

It is more difficult to talk about future biological evolution, since it is a much less deterministic process with the absence of clear boundary conditions, in comparison with geological processes. Obviously, biological evolution will be largely associated with the development of human civilization, about which it is even more difficult to assume something definite. In this process, we are both actors and spectators, and our responsibility for the future is paradoxically combined with the historical inevitability of the development of the world and society.

References

Ahrens, N. C., Jahren, A. H. (2000). Carbon isotope excursion in atmospheric CO_2 at the Cretaceous-Tertiary boundary: evidence from terrestrial sediments. *Palaios* 15(4), 314–322. https://doi.org/10.2307/3515539.

Alsos, I. G., Engelskjøn, T., Gielly, L., Taberlet, P., Brochmann, C. (2005). Impact of ice ages on circumpolar molecular diversity: insights from an ecological key species. *Molecular Ecology* 14(9), 2739–2753. https://doi.org/10.1111/j.1365-294X.2005.02621.x.

Andrews, M., Lea, P. J., Raven, J. A., Lindsey, K. (2004). Can genetic manipulation of plant nitrogen assimilation enzymes result in increased crop yield and greater N-use efficiency? An assessment. *Annals of Applied Biology* 145(1), 25–40. https://doi.org/10.1111/j.1744-7348.2004.tb00356.x.

Atkin, O. K., Millar, A. H., Gardeström, P., Day, D. A. (2000). Photosynthesis, carbohydrate metabolism and respiration in leaves of higher plants. In: Leegood, R. C., Sharkey, T. D., von Caemmerer, S (eds). *Photosynthesis: Physiology and Metabolism*. Kluwer, Dordrecht, Netherlands, pp. 153-175.

Bader, O. N., Wagner, G. K., Krieger, N. I. (1975). The role of A. A. Mansurov in the study of the Ryazan region. In: *Archeology of the Ryazan Land* (Mongait, A. L., ed.). Nauka, Moscow, pp. 324-327.

Badger, M. R., Price, G. D. (2003). CO_2 concentrating mechanisms in cyanobacteria: molecular components, their diversity and evolution. *Journal of Experimental Botany* 54(383), 609–622. https://doi.org/10.1093/jxb/erg076.

Becker, R. (1991). Compositional, nutritional, and functional evaluation of intermediate wheatgrass. *Journal of Food Processing and Preservation* 15(1), 63–77. https://doi.org/10.1111/j.1745-4549.1991.tb00154.x.

Beerling, B. J., Berner, R. A. (2000). Impact of a Permo-Carboniferous high O_2 event on the terrestrial carbon cycle. *Proceedings of the National Academy of Sciences USA* 97(23), 12428–12432. https://doi.org/10.1073/pnas.220280097.

Beerling, D.J. (2002). Low atmospheric CO_2 levels during the Permo-Carboniferous glaciation inferred from fossil lycopsids. *Proceedings of the National Academy of Sciences USA* 99(20), 12567–12571. https://doi.org/10.1073/pnas.202304999.

Beerling, D. J., Lomax, B. H., Royer, D. L., Upchurch, G. R., Kump, L. R. (2002). An atmospheric pCO_2 reconstruction across the Cretaceous-Tertiary boundary from leaf megafossils. *Proceedings of the National Academy of Sciences USA* 99(12), 7836–7840. https://doi.org/10.1073/pnas.122573099.

Beerling, D. J., Osborne, C. P., Chaloner, W. G. (2001). Evolution of leaf form in land plants linked to atmospheric CO_2 decline in the late Palaeozoic era. *Nature* 410(6826), 352–354. https://doi.org/10.1038/35066546.

Beerling, D. J., Woodward, F. I. (1997). Changes in land plant function over the Phanerozoic: reconstructions based on the fossil record. *Botanical Journal of the Linnean Society* 124(2), 137–153. https://doi.org/10.1111/j.1095-8339.1997.tb01787.x.

Bergman, M. N., Lenton, T. M., Watson, A. J. (2004). COPSE: a new model of biogeochemical cycling over Phanerozoic time. *American Journal of Science* 304(5), 397–437. https://doi.org/10.2475/ajs.304.5.397.

Berner, R. A. (1997). The rise of plants and their effect on weathering and atmospheric CO_2. *Science* 276(5312), 544–546. https://doi.org/10.1126/science.276.5312.544.

Berner, R. A. (2003). The long-term carbon cycle, fossil fuels and atmospheric composition. *Nature* 426(6964), 323–326. https://doi.org/10.1038/nature02131.

Berner, R. A., Canfield, D. E. (1989). A new model for atmospheric oxygen over Phanerozoic time. *American Journal of Science* 289(4), 333–361. https://doi.org/10.2475/ajs.289.4.333.

Berner, R. A., Kothavala, Z. (2001). GEOCARB III: a revised model of atmospheric CO_2 over Phanerozoic time. *American Journal of Science* 301(2), 182–204. https://doi.org/10.2475/ajs.301.2.182.

Bird, I. F., Cornelius, M. J., Keys, A. J. (1982). Affinity of RuBP carboxylases for carbon dioxide and inhibition of the enzymes by oxygen. *Journal of Experimental Botany* 33(5), 1004–1013. https://doi.org/10.1093/jxb/33.5.1004.

Bjerrum, C. J., Canfield, D. E. (2004). New insights into the burial history of organic carbon on the early Earth. *Geochemistry, Geophysics, Geosystems* 5(8): Q08001. https://doi.org/10.1029/2004GC000713.

Bond, W. J., Midgley, G. F., Woodward, F. I. (2003). The importance of low atmospheric CO_2 in promoting the spread of grasslands and savannas. *Global Change Biology* 9(7), 973–982. https://doi.org/10.1046/j.1365-2486.2003.00577.x.

Brooks, A., Farquhar, G. D. (1985). Effect of temperature on the CO_2/O_2 specificity of ribulose-1,5-bisphosphate carboxylase/oxygenase and the rate of respiration in light. *Planta* 165(3), 397–406. https://doi.org/10.1007/BF00392238.

Cavalier-Smith, T. (1993). Kingdom protozoa and its 18 phyla. *Microbiological Reviews* 57(4), 953–994. https://doi.org/10.1128/mr.57.4.953-994.1993.

Cen, Y.-P., Turpin, D. H., Layzell, D. B. (2001). Whole-plant gas exchange and reductive biosynthesis in white lupin. *Plant Physiology* 126(4), 1555–1565. https://doi.org/10.1104/pp.126.4.1555.

Chen, J., Montañez, I. P., Zhang, S., Isson, T. T., Macarewich, S. I., Planavsky, N. J., Zhang, F., Rauzi, S., Daviau, K., Yao, L., Qi, Y. P., Wang, Y., Fan, J. X., Poulsen, C. J., Anbar, A. D., Shen, S. Z., Wang, X. D. (2022). Marine anoxia linked to abrupt global warming during Earth's penultimate icehouse. *Proceedings of the National Academy of Sciences USA* 119(19), e2115231119. https://doi.org/10.1073/pnas.2115231119.

Cornette, J. L., Lieberman, B. S., Goldstein, R. H. (2002). Documenting a significant relationship between macroevolutionary origination rates and Phanerozoic pCO_2 levels. *Proceedings of the National Academy of Sciences USA* 99(12), 7832–7835. https://doi.org/10.1073/pnas.122225499.

Cowling, S. A. (2013). Did early land plants use carbon-concentrating mechanisms? Trends Plant Sci. 18(3), 120–124. https://doi.org/10.1016/j.tplants.2012.09.009.

Cowling, S. A., Sage, R. F. (1998). Interactive effects of low atmospheric CO_2 and elevated temperature on growth, photosynthesis and respiration in *Phaseolus vulgaris*. *Plant, Cell and Environment* 21(4), 427–435. https://doi.org/10.1046/j.1365-3040.1998.00290.x.

Crutzen, P. J. (2002). The "Anthropocene." *Journal de Physique* IV 12: 1–5. https://doi.org/10.1051/jp4:20020447.

Cuffey, K. M., Vimeux, F. (2001). Covariation of carbon dioxide and temperature from the Vostok ice core after deuterium-excess correction. *Nature* 412(6846), 523–527. https://doi.org/10.1038/35087544.

Davis, C. V., Rivest, E. B., Hill, T. M., Gaylord, B., Russell, A. D., Sanford, E. (2017). Ocean acidification compromises a planktic calcifier with implications for global carbon cycling. *Sci. Rep.* 7(1), 2225. https://doi.org/10.1038/s41598-017-01530-9.

Di Marco, G., Manes, F., Tricoli, D. Vitale, E. (1990). Fluorescence parameters measured concurrently with net photosynthesis to investigate chloroplastic CO_2 concentration in leaves of *Quercus ilex* L. *Journal of Plant Physiology* 136(5), 538–543. https://doi.org/10.1016/s0176-1617(11)80210-5.

Di Michele, W. A., Pfefferkorn, H. W., Gastaldo, R. A. (2001), Response of late Carboniferous and early Permian plant communities to climate change. *Annual Review of Earth and Planetary Science* 29, 461–487. https://doi.org/10.1146/annurev.earth.29.1.461.

Diatloff, E., Roberts, M., Saunders, D., Roberts, S. K. (2004). Characterization of anion channels in the plasma membrane of Arabidopsis epidermal root cells and the identification of a citrate permeable channel induced by phosphate starvation. *Plant Physiology* 136(4), 4136–4149. https://doi.org/10.1104/pp.104.046995.

Dolotov, Y. A. (1999). Results of a search expedition to the Kasimov district of the Ryazan region. *Spelestological Yearbook of ROSI 1999*. Moscow: ROSI-ROSS, pp. 38-57.

Donoghue, P. C. J., Harrison, C. J., Paps, J., Schneider, H. (2021). The evolutionary emergence of land plants. *Curr. Biol.* 31(19), R1281–R1298. https://doi.org/10.1016/j.cub.2021.07.038.

EPICA community members (2004) Eight glacial cycles from an Antarctic ice core. *Nature* 429(6992), 623–628. https://doi.org/10.1038/nature02599.

Evans, J. R., von Caemmerer, S. (1996). Carbon dioxide diffusion inside leaves. *Plant Physiology* 110(2), 339–346. https://doi.org/10.1104/pp.110.2.339.

Ezaki, S., Maeda, N., Kishimoto, T., Atomi, H., Imanaka, T. (1999). Presence of a structurally novel type ribulose bisphosphate carboxylase/oxygenase in the hyperthermophilic archaeon, *Pyrococcus kodakaraensis* KOD1. *Journal of Biological Chemistry* 274(8), 5078–5082. https://doi.org/10.1074/jbc.274.8.5078.

Falcon-Lang, H. J. (2006). A history of research at the Joggins Fossil Cliffs, the world's finest Pennsylvanian section. *Proceedings of the Geologists' Association* 117(3), 377-392. https://doi.org/10.1016/S0016-7878(06)80044-1.

Falkowski, P., Scholes, R. J., Boyle, E., Canadell, J., Canfield, D., Elser, J., Gruber, N., Hibbard, K., Hogberg, P., Linder, S., Mackenzie, F. T., Moore, B., Pedersen, T., Rosenthal, Y., Seitzinger, S., Smetacek, V., Steffen, W. (2000). The global carbon cycle: a test of our knowledge of Earth as a system. *Science* 290(5490), 291–296. https://doi.org/10.1126/science.290.5490.291.

Farquhar, G. D., von Caemmerer, S., Berry, J. A. (1980). A biochemical model of photosynthetic CO_2 assimilation in leaves of C_3 species. *Planta* 149(1), 78–90. https://doi.org/10.1007/BF00386231.

Field, C. B., Behrenfeld, M. J., Randerson, J. T., Falkowski, P. (1998). Primary production of the biosphere: Integrating terrestrial and oceanic components. *Science* 281(5374), 237–240. https://doi.org/10.1126/science.281.5374.237.

Gordon, R. (2024). Origin of Life via Archaea: Shaped Droplets to Archaea First, With a Compendium of Archaea Micrographs, Chapter 2. *The series: Astrobiology Perspectives on Life of the Universe*, Eds. Richard Gordon & Joseph Seckbach. Wiley-Scrivener, Beverly, Massachusetts, USA.

Gottlieb, O. R., Borin, M. R. D. B. (1998). Evolution of angiosperms via modulation of antagonisms. *Phytochemistry* 49(1), 1–15. https://doi.org/10.1016/s0031-9422(97)00843-1.

Goudie, A. (2004). Baer's Law of stream deflection. *Earth Sciences History* 23 (2), 278–282.

Grachev, A. F., Nikolaev, V. A., Nikolaev, V. G. (2006). East European platform development in the Late Precambrian and Paleozoic: Structure and sedimentation. *Russian Journal of Earth Science* 8(4), 1–22. https://doi.org/10.2205/2006ES000203.

Gunn, L. H., Martin Avila, E., Birch, R., Whitney, S. M. (2020). The dependency of red Rubisco on its cognate activase for enhancing plant photosynthesis and growth. *Proc. Natl Acad. Sci. USA* 117(41), 25890–25896. https://doi.org/10.1073/pnas.2011641117.

Harrison, T. M. (2009). The Hadean crust: evidence from >4 Ga zircons. *Annual Review of Earth and Planetary Sciences* 37(1), 479–505. https://doi.org/10.1146/annurev.earth.031208.100151.

Holland, H. D. (2006). The oxygenation of the atmosphere and oceans. *Philosophical Transactions of the Royal Society of London, Series B: Biological Sciences* 361(1470), 903–915. https://doi.org/10.1098/rstb.2006.1838.

Hönisch, B., Ridgwell, A., Schmidt, D. N., Thomas, E., Gibbs, S. J., Sluijs, A., Zeebe, R., Kump, L., Martindale, R. C., Greene, S. E., Kiessling, W., Ries, J., Zachos, J. C., Royer, D. L., Barker, S., Marchitto, T. M. Jr, Moyer, R., Pelejero, C., Ziveri, P., Foster, G. L., Williams, B. (2012). The geological record of ocean acidification. *Science* 335(6072), 1058–1063. https://doi.org/10.1126/science.1208277.

Igamberdiev, A. U. (2021). Human-driven spreading and evolution of plants during the Holocene epoch: The pioneering works of Valery Taliev. *Biosystems* 210, 104567. https://doi.org/10.1016/j.biosystems.2021.104567.

References

Igamberdiev, A. U. (2025). Human-driven evolution of cultivated plants and the origin of early civilizations: The concept of Neolithic revolution in the works of Nikolai Vavilov. *Biosystems* 247, 105359. https://doi.org/10.1016/j.biosystems.2024.105359.

Igamberdiev, A. U., Bykova, N. V. (2023). Mitochondria in photosynthetic cells: Coordinating redox control and energy balance. *Plant Physiology* 191(4), 2104–2119. https://doi.org/10.1093/plphys/kiac541.

Igamberdiev, A. U., Lea, P. J. (2002). The role of peroxisomes in the integration of metabolism and evolutionary diversity of photosynthetic organisms. *Phytochemistry* 60(7), 651–674. https://doi.org/10.1016/s0031-9422(02)00179-6.

Igamberdiev, A. U., Lea, P. J. (2006). Land plants equilibrate O_2 and CO_2 concentrations in the atmosphere. *Photosynthesis Research* 87(2), 177–194. https://doi.org/10.1007/s11120-005-8388-2.

Igamberdiev, A. U., Lea, P. J. (2006). Land plants equilibrate O_2 and CO_2 in the atmosphere. *Photosynthesis Research* 87(2), 177-194. https://doi.org/10.1007/s11120-005-8388-2.

Jones, D. L. (1998). Organic acids in the rhizosphere – a critical review. *Plant and Soil* 205(1), 25–44. https://doi.org/10.1023/A:1004356007312.

Jones, T. P. (1994). ^{13}C Enriched lower Carboniferous fossil plants from Donegal, Ireland – carbon isotope constraints on taphonomy, diagenesis and paleoenvironment. *Review of Paleobotany and Palynology* 81(1), 53–64. https://doi.org/10.1016/0034-6667(94)90126-0.

Jordan, D. B., Ogren, W. L. (1983). Species variation in kinetic properties of ribulose 1,5-bisphosphate carboxylase oxygenase. *Archives of Biochemistry and Biophysics* 227(2), 425–433. https://doi.org/10.1016/0003-9861(83)90472-1.

Karol, K. G., McCourt, R. M., Cimino, M. T., Delwiche, C. F. (2001). The closest living relatives of land plants. *Science* 294(5550), 2351–2353. https://doi.org/10.1126/science.1065156.

Kasting, J. F., Ono, S. (2006). Palaeoclimates: the first two billion years. *Philosophical Transactions of the Royal Society of London, Series B: Biological Sciences* 361(1470), 917–929. https://doi.org/10.1098/rstb.2006.1839.

Kihara, T., Wada, T., Suzuki, Y., Hara, T., Koyama, H. (2003). Alteration of citrate metabolism in cluster roots of white lupine. *Plant and Cell Physiology* 44(9), 901–908. https://doi.org/10.1093/pcp/pcg115.

Knoll, A. H. (1991). End of the Proterozoic eon. *Scientific American* 265(4), 64-73. https://doi.org/10.1038/scientificamerican1091-64.

Knoll, A. H. (1994). Proterozoic and early Cambrian protists – evidence for accelerating evolutionary tempo. *Proceedings of the National Academy of Sciences USA* 91(15), 6743–6750. https://doi.org/10.1073/pnas.91.15.6743.

Kostenchuk, N. A., Tyuryukanov, A. N. (1980). The Origin of the "Oka Flora" and Biogeocenology. *Bulletin of the MOIP, Biological Series* 85, 123-134.

Laing, W. A., Ogren, W. L., Hageman, R. H. (1974). Regulation of soybean net photosynthetic CO_2 fixation, by interaction of CO_2, O_2 and ribulose 1,5-diphosphate carboxylase. *Plant Physiology* 54(5), 678–685. https://doi.org/10.1104/pp.54.5.678.

Lenton, T. M. (1998). Gaia and natural selection. *Nature* 394(6692), 439–447. https://doi.org/10.1038/28792.

Lenton, T. M. (2001). The role of land plants, phosphorus weathering and fire in the rise and regulation of atmospheric oxygen. *Global Change Biology* 7(6), 613–629. https://doi.org/10.1046/j.1354-1013.2001.00429.x.

Lenton, T. M., Dahl, T. W., Daines, S. J., Mills, B. J., Ozaki, K., Saltzman, M. R., Porada, P. (2016). Earliest land plants created modern levels of atmospheric oxygen. *Proceedings of the National Academy of Sciences of the United States of America* 113(35), 9704–9709. https://doi.org/10.1073/pnas.1604787113.

Lenton, T. M., Watson, A. J. (2004). Biotic enhancement of weathering, atmospheric oxygen and carbon dioxide in the Neoproterozoic. *Geophysical Research Letters* 31(5), L05202. https://doi.org/10.1029/2003GL018802.

Lyu, H., Yim, W. C., Yu, Q. (2024) Genomic and transcriptomic insights into the evolution of C4 photosynthesis in grasses. *Genome Biol. Evol.* 16(8), https://doi.org/10.1093/gbe/evae163.

Lyubishchev, A. A. (1982). *Problems of the Form, Systematics, and Evolution of Organisms.* Nauka, Moscow.

Mahelka, V., Kopecký, D., Paštová, L. (2011) On the genome constitution and evolution of intermediate wheatgrass (*Thinopyrum intermedium*: Poaceae, Triticeae). *BMC Evolutionary Biology* 11 127. https://doi.org/10.1186/1471-2148-11-127.

Makarova, N. V., Makarov, V. I., Geptner, T. M., Sukhanova, T. V. (1999). The latest tectonics of the Oka-Tsninsky uplift. *Gerald of Moscow University, Series 4, Geology* 4, 22-28.

Matthaeus, W. J., Macarewich, S. I., Richey, J. D., Wilson, J. P., McElwain, J. C., Montañez, I. P., DiMichele, W. A., Hren, M. T., Poulsen, C. J., White, J. D. (2021). Freeze tolerance influenced forest cover and hydrology during the Pennsylvanian. *Proceedings of the National Academy of Sciences USA* 118(42), e2025227118. https://doi.org/10.1073/pnas.2025227118.

Maurice, M., Tosi, N., Schwinger, S., Breuer, D., Kleine, T. (2020). A long-lived magma ocean on a young Moon. *Science Advances* 6(28), eaba8949. https://doi.org/10.1126/sciadv.aba8949.

McCarthy, D. D., Seidelmann, K. P. (2009). *Time: From Earth Rotation to Atomic Physics*. John Wiley & Sons, NJ, USA. ISBN 978-3-527-62795-0.

McElwain, J. C., Beerling, D. J., Woodward, F. I. (1999). Fossil plants and global warming at the Triassic-Jurassic boundary. *Science* 285(5432), 1386–1390. https://doi.org/10.1126/science.285.5432.1386.

McIlroy, D., Dufour, S.C., Taylor, R., Nicholls, R. (2021). The role of symbiosis in the first colonization of the seafloor by macrobiota: Insights from the oldest Ediacaran biota (Newfoundland, Canada). *Biosystems* 205, 104413. https://doi.org/10.1016/j.biosystems.2021.104413.

Menning, M., Alekseev, A. S., Chuvashov, B. I., Davydov, V. I., Devuyst, F.-X., Forke, H. C., Grunt, T. A., Hance, L., Heckel, P. H., Izokh, N. G., Jin, Y.-G., Jones, P. J., Kotlyar, G. V., Kozur, H. W., Nemyrovska, T. I., Schneider, J. W., Wang, X.-D., Weddige, K., Weyer, D., Work, D. M. (2006). Global time scale and regional stratigraphic reference scales of Central and West Europe, East Europe, Tethys, South

China, and North America as used in the Devonian–Carboniferous–Permian Correlation Chart 2003 (DCP 2003). *Palaeogeography, Palaeoclimatology, Palaeoecology* 240(1-2), 318–372. https://doi.org/10.1016/j.palaeo.2006.03.058.

Meyen, S. V. (1987). Geography of macroevolution in higher plants. *Zhurnal Obshchei Bilogii [Journal of General Biology]* 48(3), 291-309.

Mongait, A. L., ed. (1975). *Archaeology of Ryazan Land*. Nauka, Moscow.

Morrison, L. A. (2004) CEREALS | Evolution of Species. In: Wrigley, C. (Editor), *Encyclopedia of Grain Science*, pp. 262-273. Elsevier, Amsterdam. https://doi.org/10.1016/B0-12-765490-9/00049-5.

Moulton, K. L., West, J., Berner, R. A. (2000). Solute flux and mineral mass balance approaches to the quantification of plant effects on silicate weathering. *American Journal of Science* 300(7), 539–570. https://doi.org/10.2475/ajs.300.7.539.

Ossa, F. O., Spangenberg, J. E., Bekker, A., König, S., Stüeken, E. E., Hofmann, A., Poulton, S. W., Yierpan, A., Varas-Reus, M. I., Eickmann, B., Andersen, M. B., Schoenberg, R. (2022). Moderate levels of oxygenation during the late stage of Earth's Great Oxidation Event. *Earth and Planetary Science Letters* 594, 117716. https://doi.org/10.1016/j.epsl.2022.117716.

Pawlowski, J., Holzmann, M., Berney, C., Fahrni, J., Gooday, A. J., Cedhagen, T., Habura, A., Bowser, S. S. (2003). The evolution of early Foraminifera. *Proceedings of the National Academy of Sciences USA* 100 (20), 11494–11498. https://doi.org/10.1073/pnas.2035132100.

Pearson, P. N., Ditchfield, P. W., Singano, J., Harcourt-Brown, K. G., Nicholas, C. J., Olsson, R. K., Shackleton, N. J., Hall, M. A. (2001). Warm tropical sea surface temperatures in the Late Cretaceous and Eocene epochs. *Nature* 413(6855), 481–487. https://doi.org/10.1038/35097000.

Raven, J. A. (1991). Plant response to high O_2 concentrations – relevance to previous high O_2 episodes. *Global and Planetary Change* 97(1-2), 19–38. https://doi.org/10.1016/0921-8181(91)90124-F.

Raven, J. A. (2003). Inorganic carbon concentrating mechanisms in relation to the biology of algae. *Photosynthesis Research* 77(2-3), 155–171. https://doi.org/10.1023/A:1025877902752.

Raven, J. A., Falkowski, P. G. (1999). Oceanic sinks for atmospheric CO_2. *Plant, Cell and Environment* 22(6), 741–755. https://doi.org/10.1046/J.1365-3040.1999.00419.X.

Retallack, G. J. (2001). A 300-million-year record of atmospheric carbon dioxide from fossil plant cuticles. *Nature* 411(6835), 287–300. https://doi.org/10.1038/35077041.

Rial, J. (2004). Abrupt climate change: chaos and order at orbital and millennial scales. *Global and Planetary Change* 41(2), 95–109. https://doi.org/10.1016/j.gloplacha.2003.10.004.

Richey, J. D., Montañez, I. P., Goddéris, Y., Looy, C. V., Griffis, N. P., DiMichele, W. A. (2020). Influence of temporally varying weatherability on CO_2-climate coupling and ecosystem change in the late Paleozoic. *Climate of the Past* 16, 1759–1775. https://doi.org/10.5194/cp-16-1759-2020.

Rogers, J. J., Santosh, M. (2004). *Continents and supercontinents*. Oxford University Press, Oxford. https://doi.org/10.1016/S1342-937X(05)70827-3.

Rothman, D. H. (2001). Global biodiversity and the ancient carbon cycle. *Proceedings of the National Academy of Sciences USA* 98(8), 4305–4310. https://doi.org/10.1073/pnas.071047798.

Royer, D. L., Berner, R. A., Beerling, D. J. (2001) Phanerozoic atmospheric CO_2 change: evaluating geochemical and paleobiological approaches. *Earth Science Reviews* 54(4), 349–392. https://doi.org/10.1016/S0012-8252(00)00042-8.

Sage, R. F. (2004). The evolution of C_4 photosynthesis. *New Phytologist* 161(2), 341–370. https://doi.org/10.1111/j.1469-8137.2004.00974.x.

Sage, R. F., Coleman, J. R. (2001). Effects of low atmospheric CO_2 on plants: more than a thing of the past. *Trends in Plant Science* 6(1), 18–24. https://doi.org/10.1016/s1360-1385(00)01813-6.

Sage, R. F., Kubien, D. S. (2003). Quo vadis C_4? An ecophysiological perspective on global change and the future of C_4 plants. *Photosynthesis Research* 77(2-3), 209–225. https://doi.org/10.1023/A:1025882003661.

Sahney, S., Benton, M. J., Falcon-Lang, H. J. (2010). Rainforest collapse triggered Pennsylvanian tetrapod diversification in Euramerica. *Geology* 38(12), 1079–1082. https://doi.org/10.1130/G31182.1.

Savard, L., Li, P., Strauss, S. H., Chase, M. W., Michaud, M., Bousuett, J. (1994). Chloroplast and nuclear gene sequences indicate Late Pennsylvanian time for the last common ancestor of extant seed plants. *Proceedings of the National Academy of Sciences USA* 91(11), 5163–5167. https://doi.org/10.1073/pnas.91.11.5163.

Schreiber, M., Himmelbach, A., Börner, A., Mascher, M. (2018). Genetic diversity and relationship between domesticated rye and its wild relatives as revealed through genotyping-by-sequencing. *Evolutionary Applications* 12(1), 66–77. https://doi.org/10.1111/eva.12624.

Sharkey, T. D. (1988). Estimating the rate of photorespiration in leaves. *Physiologia Plantarum* 73(1), 147–152. https://doi.org/10.1111/j.1399-3054.1988.tb09205.x.

Sharov A. A., Igamberdiev A. U. (2014). Inferring directions of evolution from patterns of variation: the legacy of Sergei Meyen. *Biosystems* 123, 67–73. http://dx.doi.org/10.1016/j.biosystems.2014.06.008.

Shields-Zhou, G., Och, L. (2011). The case for a Neoproterozoic Oxygenation Event: Geochemical evidence and biological consequences. *GSA Today* 21(3), 4–11. https://doi.org/10.1130/GSATG102A.1.

Sigman, D. M., Boyle, E. A. (2001). Glacial/interglacial variations in atmospheric carbon dioxide. *Nature* 407(6806), 859–869. https://doi.org/10.1038/35038000.

Tajika, E. (2003). Faint young Sun and the carbon cycle: Implication for the Proterozoic global glaciations. *Earth and Planetary Science Letters* 214(4), 443–453. https://doi.org/10.1016/S0012-821X(03)00396-0.

Taliev, V. I. (1895). Chalk Forests of Donets and Volga Basins. *Proceedings of the Society of Naturalists at Kharkiv University* 29, 225-282.

Taliev, V. I. (1896). An outline of weed biology. *Natural Science and Geography* 8, 812-821.

Tarduno, J. A. (2025). Earth's magnetic dipole collapses, and life explodes. *Physics Today* 78(4), 26–33. https://doi.org/10.1063/pt.bide.yfhb.

Tolbert, N. E., Benker, C., Beck, E. (1995). The oxygen and carbon dioxide compensation points of C_3 plants: possible role in regulating atmospheric oxygen. *Proceedings of the National Academy of Science USA* 92(24), 11230–11233. https://doi.org/10.1073/pnas.92.24.11230.

Tortell, P. D. (2000). Evolutionary and ecological perspectives on carbon acquisition in phytoplankton. *Limnology and Oceanography* 45(3), 744–750. https://doi.org/10.4319/lo.2000.45.3.0744.

Uemura, K., Anwaruzzaman, Miyachi, S., Yokota, A. (1997). Ribulose-1,5-bisphosphate carboxylase/oxygenase from thermophilic red algae with a strong specificity for CO_2 fixation. *Biochemical and Biophysical Research Communications* 233(2), 568–571. https://doi.org/10.1006/bbrc.1997.6497.

Vavilov, N. (1926). *Origin and Geography of Cultivated Plants*. Cambridge Univ. Press, Cambridge, 1992.

Vernadsky, V. I. (1926). *Biosphere*. Copernicus Books, Springer, New York, 1998.

von Bloh, W., Bounama, C., Franck, S. (2003). Cambrian explosion triggered by geosphere-biosphere feedbacks. *Geophysical Research Letters* 30(18), 1963. https://doi.org/10.1029/2003GL017928.

von Caemmerer, S., Furbank, R.T. (2003). The C_4 pathway: an efficient CO_2 pump. *Photosynthesis Research* 77(2-3), 191–207. https://doi.org/10.1023/A:1025830019591.

Wagoner, P., Schauer, A. (1990). Intermediate wheatgrass as a perennial grain crop. In: Janick, J. and Simon, J. E. (eds.), *Advances in new crops*. Timber Press, Portland, OR, pp. 143-145.

Whitney, S. M., Baldett, P., Hudson, G. S., Andrews, T. J. (2001). Form I Rubiscos from non-green algae are expressed abundantly but not assembled in tobacco chloroplasts. *Plant Journal* 26(5), 535–547. https://doi.org/10.1046/j.1365-313x.2001.01056.x.

Wildman, R. A., Hickey, L. J., Dickinson, M. B., Berner, R. A., Robinson, J. M., Dietrich, M., Essenhigh, R. H., Wildman, C. B. (2004). Burning of forest materials under late Paleozoic high atmospheric oxygen levels. *Geology* 3(5), 457–460. https://doi.org/10.1130/G20255.1.

Williams, R. J. P. (2011). Chemical advances in evolution by and changes in use of space during time. *Journal of Theoretical Biology* 268(1), 146–159. https://doi.org/10.1016/j.jtbi.2010.09.021.

Wilson, J. P., Montañez, I. P., White, J. D., DiMichele, W. A., McElwain, J. C., Poulsen, C.J., Hren, M. T. (2017). Dynamic Carboniferous tropical forests: new views of plant function and potential for physiological forcing of climate. *New Phytologist* 215(4), 1333–1353. https://doi.org/10.1111/nph.14700.

Zachos, J., Pagani, M., Sloan, L., Thomas, E., Billups, K. (2001). Trends, rhythms, and aberrations in global climate 65 Ma to present. *Science* 292(5517), 686–693. https://doi.org/10.1126/science.1059412.

Zagoskin, M. N. (1902). Kuzma Roshchin. *Historical novel*. M. V. Klyukin, Moscow. Originally published in 1836.

Zavaleta, E. S., Shaw, M. R., Chiariello, N. R., Mooney, H. A., Field, C. B. (2003) Additive effects of simulated climate changes, elevated CO_2, and nitrogen deposition on grassland diversity. *Proceedings of the National Academy of Sciences USA* 100(13), 7650–7654. https://doi.org/10.1073/pnas.0932734100.

Zegers, T. E., de Wit, M. J., White, S. H. (1998). Vaalbara, Earth's oldest assembled continent? A combined structural, geochronological, and palaeomagnetic test. *Terra Nova* 10(5), 250–259. https://doi.org/10.1046/j.1365-3121.1998.00199.x.

Appendix I: Fyodor Dashkov: History of the Village of Pervo

This material is based on the text written by my grandfather Fyodor Dashkov in the 1970s. I added two introductory paragraphs and slightly edited the text.

The village of Pervo is the first Slavic settlement along the Oka near the confluence of the Tashenka River, where a triangle is formed: on the one side is the Oka, on the other Tashenka, and on the third the forest. This triangle, which represented relative security for those living in it, was densely populated; there were villages: Pervo, Vasilyovo, Pozdnyakovo, Davydovo, and Zhdanovo. The names themselves show their Slavic origin, while outside the triangle there are Finno-Ugric names: Chinur, Yerakhtur. In the village of Zhdanovo, the peasants were state-owned and did not belong to a landowner, and this was noticeable in their more independent and freedom-loving behavior during the years of Soviet power.

The village of Pervo is located on the right bank of the Oka, 12 kilometers below the city of Kasimov. The former name of the village was Perya, and even earlier, Perenya (Perynya). The name may come from the name of the Slavic deity Perun. Probably, here, the pagans once had a sacred place where there was an idol depicting Perun and where prayers were held to this idol.

There is also a version that the name Perya (translated as "feathers") was given to the village since, in the old days, there were many birds on the banks of the river in this place. The abundance of birds is also evidenced by the name Gusinka (related to "geese") in the flood meadows on the opposite bank. Large numbers of feathers accumulated from birds in this place. At the beginning of the twentieth century, local landowners changed the name of the village to "Pervi," and after the events of October 1917, the village became known as "Pervo." The natural conditions of this village, including the proximity of fields, meadows, forests, water, and rugged terrain with deep ravines, contributed to the people who settled there to find a livelihood and engage in agriculture, hunting animals, fishing, and picking berries and mushrooms.

They could hide there from animals and the incursions of other tribes and peoples.

On the descent to the river on an elevated place, currently called "Paltso," there was an ancient settlement. Along the course of the Oka between Pervo and Kasimov, there are Baishevsky, Tashenskoye, and Popovskoye Neolithic settlements.

The village of Perya was first mentioned in 1628 in the scribal books of Voeikov as having two churches: the Nativity of the Blessed Virgin Mary and the church of Paraskeva Pyatnitsa. According to the salary books of 1676, in the village of Perya, there is a church of the Great Wonderworker Nicholas, two courtyards of priests, five courtyards of landlords, twenty households of peasants, and three bachelors (bobyl) courtyards. In the 18th century, two churches were mentioned in the village, a wooden church in honor of the Nativity of the Blessed Virgin Mary and a stone church in honor of the Nativity of the Blessed Virgin Mary with a St. Nicholas chapel. Since 1861, there was a zemstvo school in the village. In 1905, the village was part of the Betinsky parish of the Kasimov district and had 87 households with a population of 667 people.

The village has preserved many ancient monuments dating back to the times of serfdom. The leading place among them is occupied by the Old Church of the Nativity of the Blessed Virgin Mary with the St. Nicholas chapel (Figure 20). Unfortunately, the chapel was destroyed during the restoration. According to experts, it is an architectural monument of the 18th (or even the 17th) century. The bell tower was added later. Architecturally, the bell tower is a high tower with an original completion and a belfry at the level of the third, open tier. Of great interest are the crosses above the altar of the main building and the chapel in the name of St. Nicholas the Wonderworker. There is a legend that the village belonged to the princes of Shuisky and, probably, it was under them that the church was built. On the wall inside the church was an inscription indicating that Major Zasetsky, who owned the village after Shuisky, was buried there. Around the church, there was a cemetery with a large number of rough stones. There is evidence that opposite the village of Pervo on the left bank of the Oka, flood meadows belonged to the famous military commander Alexander Suvorov.

Appendix I: Fyodor Dashkov: History of the Village of Pervo

Figure 20. Architectural monuments of the village of Pervo. Upper left – Church of the Nativity of the Blessed Virgin Mary with the St. Nicholas chapel (17th-18th century), Upper right – Church of St. Diomedes (second half of the 19th century), bottom left – carriage house (19th century), bottom right – manor house and storeroom with turret (19th century). Author's photos (1975, Diomedes church – 2013). Author's photos.

In 1955, students of the Pervo School found a small jar with silver coins from the time of Tsar Alexei Mikhailovich (reigned 1645-1676). The find, which testified to the village's centuries-long existence, was handed over to the Kasimov Museum of Local History. In addition to the old church, Pervo has preserved a large number of various premises and remains of former buildings and earthworks built during the time of serfdom. The most ancient of these include the foundation with a basement of the former house of Major Zasetsky, the remains of a brick mill, and a basement, which has long been called a prison. It was a place of imprisonment for guilty serfs. At least, the layout of the basement itself speaks in favor of this version.

Later buildings include the manor house, built either by Hildebrandt or his predecessor Kiselyov, a carriage house, two storerooms with turrets (these buildings were built in the style of classicism), a barnyard, a manor (the Barskaya) road, a pond, remains of brick factories and the New Church, named after Diomedes (Figure 20). In addition, many ditches and remains of brick and stone walls have been preserved. Already in our time, some of the buildings have undergone reconstruction. So, for example, a two-story manor house became half one-story.

A very important structure is the so-called Barskaya road along the mountain slope. Its length is about 800 meters, and its width is 7-8 meters. In order to secure the embankment and protect the road from destruction, it was all planted with birch and ash trees. Obviously, a lot of labor and time of serfs was spent on the construction of the road. There is no doubt that this work was done manually.

All buildings made by serf masters are distinguished by quality and reliability. This is especially felt in the architecture of the New Church. The precision in laying bricks along horizontal and vertical lines and the elegant processing of limestone are astonishing. High skill was also shown during the construction of the carriage yard. It is a pity that the original architecture was damaged during its restoration. Thus, the dome covering the entire structure was removed, and chiseled pillars and stones were damaged. But the gates, distinguished by their power, were preserved in their original form.

There is a legend about how the New Church was built. It is known that the landowner Kiselyov had the name Diomede. He was very rich but in poor health and was often ill for a long time. His wife, a young and beautiful woman, invited a doctor named Hildebrandt to treat her husband. After Kiselyov's death, the widow married a doctor, thus making him rich and noble. To calm their conscience, the couple donated 200 thousand rubles (a gigantic amount at that time) for the construction of the "Temple of God in honor of St. Diomedes." As is often the case, the construction was delayed. In addition, part of the money was lost in card gambling, and then, to end the work, the couple sent old people all over the country to collect the missing amount. In total, the temple was built for about 50 years, and the construction was completed in 1900.

When starting to lay the foundation of the temple in honor of St. Diomedes, the Hildebrandt couple decided to make August 16 (old style), when the Orthodox Church celebrates this great martyr, a holiday for their people. They wanted the people to pray for their masters on this day. But it turned out that there were already plenty of temple holidays: in honor of the Virgin, in honor of St. Nicholas, twice a year in honor of Paraskeva the Martyr. And then the couple tried to consign the holiday in honor of the latter to oblivion, mentioning the fact that the temple named after her had long burned down. Therefore, they reasoned, it is necessary to replace this holiday with a holiday in honor of St. Diomedes. For this purpose, the burgomaster was given the task of turning over jugs of homebrew from those who were preparing it for Paraskeva's day. The people on this day were ordered to work. Indeed, in the village of Pervo, the feast in honor of Paraskeva was abolished, although in other settlements of the church parish (Davydovo, Savino), it could not be canceled. This is explained by the fact that these two villages belonged to the landowner Shuvalov, who did not care about the ideas of his neighbors.

Anyway, the Pervo peasants did not accept the holiday in honor of Diomedes. "This holiday is yours, not ours," the old men firmly declared to their masters, and on the day of the martyr Diomedes, they worked in the field.

Around the village, there are large deposits of limestone. Since ancient times, the villagers have been engaged in its extraction. Limestone was used in the construction of stone buildings. The stonemasons dug it up and sold it to buyers during the winter. In the 19th century, the main buyers of limestone were the residents of the Elatomsky district, Kugushev, and the Vasilyov brothers. In turn, the buyers resold the stone to Nizhny Novgorod.

Seasonal work was widespread in the village of Pervo. Most middle-aged and young men did not stay in the village for the summer but went to the cities to be hired to build houses and factories. Pervian women have traditionally been excellent craftswomen in the manufacture of fine yarn, linen, and cloth. Among them were many well-known embroiderers in the county. A wide range of colors distinguished embroidery on shirts.

Appendix II: From My Childhood Memories

From my early childhood, I developed an interest in natural science and music, largely thanks to my grandparents, Fyodor Andreyevich Dashkov (1900-1983) and Maria Alexandrovna Dashkova (1904-1980). Music revealed the flow of subjective time, and the surrounding nature revealed the flow of objective time. With my grandfather playing the guitar, I sang songs, mostly revolutionary and historical, but we did not sing lyrical or modern songs. When I was about four years old, under the impression of singing the ballad on the text of Mikhail Lermontov's poem "Airship, From Zedlitz," when my grandfather showed me the grave of his father, Andrei Nikolaevich Dashkov, I recited this poem: "A heavy stone lies on him, so that he could not rise from the coffin." When I was five years old, my grandfather bought me a piano, and I began to study music. My grandmother, who knew classical music very well and played piano, was my mentor.

My grandfather was the director of the local school and was one of the organizers (along with Alexei Mansurov, Ivan Kitaytsev, and Alexander Olenin) of the Kasimov Museum of Local History in the 1920s. In his philosophical thoughts, he emphasized the need to develop a single pantheistic worldview based on rational reasoning, which would be acceptable to all of humanity. One of his favorite works was Gavriil Derzhavin's ode "God." He also spoke about the need to adopt unified global principles for the rational use of the Earth's natural resources. Grandmother Maria Alexandrovna Dashkova, née Tsvetaeva, taught German and had a deep knowledge of Russian literature and classical music. Her father, Alexander Petrovich Tsvetaev (1876-1930), was a priest in the village of Alpatyevo (now Moscow Region). He was repressed in Stalin's time and died in exile. Her mother, Olga Alexeevna (1883-1972), after her husband's death, lived with her daughter and son-in-law in the village of Pervo.

In Pervo, I spent my early childhood, from one to seven years old, and then visited this place for three summer months from the town of Voronezh, where I lived with my mother and attended primary and secondary school. My grandparents had a good garden with apple trees, cherries, plums, and one pear tree (seedless variety), on which sweet fruits ripened. There were different varieties of apple trees – summer, autumn, and winter varieties. Grandfather planted the garden after the Second World War. He planted several varieties of Ivan Michurin's selection. Michurin was a famous practitioner of plant breeding. I remember the winter variety, obtained by vegetative hybridization with a pear. A small "pear" elongation of the fruit at the apple stem testified to its origin. This apple tree developed from a small branch, which was the only one left after a very cold post-war winter. We loved the summer Michurin variety Golden Kitaika. These apples ripened earlier than all others, were juicy, and tasty. Grandmother called another variety of juicy apples "Favorite." I do not know what variety it was, but the apples were very tasty. By the beginning of July, Vladimirskaya cherries ripened; they were sweet and dark. Another cherry, Kolomenskaya, ripened on tall trees later and was more sour. And at the edges of the garden, my grandfather planted the Michurin variety, the Beauty of the North. This cherry was shorter, more like a bush; the fruits were light red, and the pit was difficult to separate from the stalk. At one time, my grandfather thought that the Michurin variety could not be that bad and that it was actually a steppe shrub cherry. Most likely, a Michurin variety did not have sufficient genetic stability and had lost its best properties, if there were any. In any case, this cherry was similar to the one depicted in the color illustration of a large book describing Michurin varieties. I read this book with great interest.

The landmark was a large elm tree near the house. It had a large crown with dense foliage. One summer, the leaves began to thin out, the tree turned out to be sick, and it dried up in a few years. Impressed by conversations about environmental pollution, I began to think that this was the reason for the death of the elm and even wrote a poem in the style of Alexei Koltsov's poetry, dedicated to this tree. The real reason was Dutch elm disease (*elm graphiosis*), a fungal disease transmitted by bark beetles and other insects, which was a fairly common phenomenon. The disease was presumably brought from East Asia to Europe at the beginning of the 20th century and was first described in the Netherlands.

Alleys of white and pink lilacs survived by the time of my childhood from the garden of the old manor house of the former owner of this place (landlord), Andrei Ivanovich Hildebrandt. The steep slope (called Barskaya Gora) was planted with linden trees, where I found columbines (*Aquilegia*), which turned out to be the unpretentious plants that survived on the Oka slope. Most were blue, but there were also purple, yellow, and white ones. I transplanted the

Appendix II: From My Childhood Memories

columbines of different colors into my grandfather' garden. Then I found many wild lupines on the site of Andrei Ivanovich's brother's estate near the village of Betino, on the steep bank of the Oka. These flowers turned out to be the wild descendants of witnesses to the recent social cataclysm. Of course, the terrible memories of the war were still fresh. Grandfather and his three brothers (on his father's side) went through the entire war and all returned. But grandfather's cousins, the Terekhovs (five brothers), all died. Two of my grandmother's brothers died. I never played with war toys, but someone once gave me a toy gun. I was about four years old then. I immediately threw it into a barrel of rainwater. Grandfather's brother Vasily Andreevich, who arrived in Pervo that day, said: "This is how wonderfully the problem of disarmament has been solved." Alexander Filippovich Trukhachev, who worked as an accountant on a collective farm, was left without hands and feet. He was given a small disabled car in which he rode. I remember him quite well; he was a friendly and open person. Another war participant whom I remember is Alexei Mikhailovich Shestopalov, a mathematics teacher from the village of Betino. He was a cheerful, kind person who liked to drink vodka and was a friend of my grandfather and grandmother. I remember how he said that at the war, he no longer reacted to flying bullets, only to exploding shells. He said that he never aimed to shoot people in the war, and if others wanted to kill him, that was their problem. Humanistic ethics was in his soul, although he never uttered the word "sin" and had no religious beliefs. Once he rode his bicycle to Pervo and invited my grandfather to drink with him, and my grandfather answered him very definitely that he would not drink, since he saw that Alexei Mikhailovich had already drunk enough. Then Alexei Mikhailovich, after several unsuccessful persuading attempts, went to the chairman of the collective farm, who was his distant relative. After they drank some more there, Alexei Mikhailovich rode his bicycle home to the village of Betino, six kilometers away. On the way, he apparently had a stroke, fell off his bicycle, and died.

And there was another story. One man, also from the village of Betino, visited my grandfather to show his war memoirs. Grandfather read them and asked how it happened that, as written in this text, everyone died during the breakout from the encirclement, but this man remained alive. He answered something unintelligible and then said that he would burn all his papers on the stove, which was done. After that, he communicated little, and after some time, he died.

Grandfather actively participated in the public life of the village even after retirement. He fought against the abuses of the administration and defended ordinary workers. The administration of the collective farm was actively engaged in falsifications, so, as grandfather found out, for one job of harvesting hay, workers were issued 13 rubles, signatures were collected, and then 1 was changed to 4, and the collective farm chairman and his assistants put the difference in their pockets. Grandfather actively fought against this. Later, the next chairman staged a fire in his house and moved into a luxurious house that had been built, appropriating it for himself. Grandfather actively opposed this and went to the secretary of the Kasimov city party committee (who, as it turned out later, had also appropriated the property). As a result, the property was returned from the chairman in favor of the collective farm, but the chairman's residence in this new house was registered as renting an apartment.

In addition to the main village cemetery, Pervo had a small cemetery for the local "elite," called the Bratskoye (Brotherhood) Cemetery, where, among others, those who had once created collective farms were buried. It is located in the center of the village, near the church of St. Diomedes. Now it is completely overgrown and almost forgotten. When I was little, my grandfather once said that perhaps he would be buried in this cemetery, since he was the director of the local school and actively participated in public life. He never spoke about this again. Later I learned that one of the villagers said to him, pointing to the Bratskoye Cemetery: "Every day I ride on my horse past those buried here and think: may you be damned." Grandfather often recalled his friend from his youth, Pyotr Drozdov, who became a professor of history and editor of the journal "The Marxist Historian." Pyotr Drozdov became a victim of Stalin's Great Terror in 1937. Grandfather thought that he had died somewhere in Kolyma, but it turned out that he was executed in Moscow, on the territory of the Donskoy Monastery.

The time of my childhood corresponded to the period of global cooling, which lasted from the 1940s to the 1970s and ended with the hot summer of 1972. I remember that the papers in popular science journals of that time suggested that human activity leads to a cooling of the climate on Earth due to the release of aerosols into the atmosphere. When this period ended, the main trend became the justification of global warming. Often, when I came to Pervo in late May–early June in my childhood, it was rainy and cold.

Near the village of Pervo, there is a large ravine, called by local residents Guzhovka. I have one memory associated with a deep side ravine, branching off from the main ravine, the longest one. Perhaps a person who escaped from prison was hiding in it. There was a lot of talk about this when I was about 12 years old. Some people saw this man; he even seemed to be chasing someone, and someone was passing him food. A couple of years later, going down into this ravine, I found an abandoned hut, carefully camouflaged from the side of the village of Shemyakino. I also remember how we observed the solar eclipse on May 20, 1966. Through the smoked glass, we saw the Sun, which was covered by the Moon by seventy percent.

This was the time of the first flights into space. I do not remember the flight of Yuri Gagarin, I was less than two years old, but I remember a little about how people talked about the flight of Valentina Tereshkova, and then Vladimir Komarov, Konstantin Feoktistov, and Boris Egorov, about Alexei Leonov's spacewalk. Everyone grieved when Komarov crashed on Soyuz 1. While on vacation in Pervo, I saw Neil Armstrong and Edwin Aldrin land on the Moon on July 20, 1969. We would go to a neighbor's house to watch TV (my grandparents did not have a TV back then). In Voronezh, an elementary school teacher would write the names of new Soviet cosmonauts on the board, and when a message came in about the flight of American astronauts, I wrote their names on the board during the break between classes, which the teacher immediately erased.

My grandparents were very worried, watching how the village was degrading, how the rural population was drinking itself to death. My grandfather, however, kept the revolutionary ideas of his youth within himself, believing that Stalinism was a perversion of these ideas. My grandmother had a rather negative attitude towards the revolution, but tried not to discuss this topic. She recalled that before the First World War, the country was developing rapidly, and in her village of Alpatyevo, people said that soon there would not be a single house with a thatched roof left. There were periods of hope: in the 1960s during the "thaw" when my grandfather managed to organize an amateur theater in the village, where they staged Alexander Ostrovsky, and in the early 1970s, during the "détente." My grandparents listened to the foreign radio (mainly BBC) through jammers, and during the "détente," the jamming was less. I remember the announcement on BBC radio: "Our political observer Anatoly Maximovich Goldberg is at the microphone." Anatoly Maximovich was remarkable in that he revealed the human face of the Western world to me, and it was not so important how profound his comments were, although he always demonstrated a high professional level of political analysis. His pleasant baritone and intelligent communication attracted many listeners. I remember how, already in Voronezh, I quoted in a circle of acquaintances his statement about the senselessness of further multiplication of the nuclear arsenal. "Anatoly Maximovich – who is this, your teacher?" – I was asked. In a sense, he was my teacher. Already at the age of

50, I walked past the building (the Bush House) in London, from which Anatoly Maximovich Goldberg had a broadcast in the 1960s-1970s.

When I arrived in Voronezh, I was surprised that the university professors I spoke with were much more loyal to the authorities as compared to my grandparents, despite their high level of education and academic degrees. It did not occur to me then that they already subconsciously considered themselves obliged to this government for their rather high social rank at that time. The time print of my early childhood is best represented in photographs from that time. New photographs do not give the same idea, despite their better quality. Space has not changed so much, and time, that elusive thing that is difficult to express in words, has gone; the "chronotope" has changed, to use the words of the thinker Mikhail Bakhtin. I came to these places in the summer of 1960, when I was one year old, and three years before that the film "The Road to Calvary" was released, where the heroes sail not along the Volga as in the original text, but along the Oka past the Church of Elijah the Prophet in Kasimov's old town. I look closely at this frame and see that this is the Oka of my childhood...

Appendix III: Issues of the Oka Vegetation Distribution: The Oka Flora in the Kasimov Region of the Ryazan Oblast

This is the author's first paper (Figure 21), completed in September 1974. It was preceded by the description of some plants of the Oka flora in 1973. The original text is preserved with only minor editing.

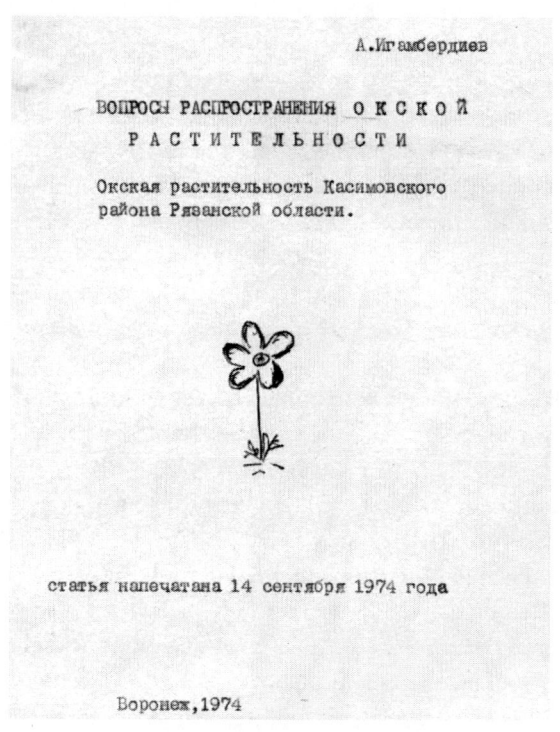

Figure 21. Printed title page of the author's paper of 1974.

Nikolai Kaufman was one of the first to draw attention to the peculiarities of the Oka flora. He divided the flora of the Moscow province into the Oka flora and the proper Moscow flora. Kaufman noticed the appearance along the Oka of several dozens of species growing in abundance to the south, in the steppes and shrubs of the chernozem (black soil) strip. He suggested that the currents of the Oka brought the seeds of these plants that settled on its banks (Kaufman, 1866).

Later, Dmitry Litvinov put forward his ideas on the origin of the Oka vegetation. In the article "On the Oka Flora" (Litvinov, 1899), he explained the appearance of southern plants on the Oka by the relict hypothesis. He considered the plants of the limestone slopes to be relics of the pre-glacial period, which were preserved on the rocky outcrops of riverbanks that had not undergone changes in either relief or physical and chemical properties (Litvinov, 1902). Litvinov believed that before the glacier, steppe vegetation was widespread far to the north.

Litvinov's relic hypothesis is now accepted by most botanists and biogeographers. But there are other points of view. Valery Taliev completely rejected Litvinov's concept. He considered both the vegetation of chalk, sandy, and limestone outcrops, and the weeds to be alien as a result of human activity (Taliev, 1895). He pointed out that chalk, sand, and limestone outcrops were created by human and animal activity, which destroyed the upper layers of soil. Remnants of pine forests are preserved on these outcrops, and unpretentious plants settle near them.

Of course, not all the conclusions of Taliev can be accepted, but it is necessary to point out the advantage of his theory over the relic hypothesis of Litvinov.

The current study of the Oka vegetation was performed in the Kasimov district of the Ryazan region. This area spans from the city of Kasimov, 30 km down the Oka to the village of Balushevo-Pochinki, and along the Tashenka River from its confluence with the Oka to the village of Zhdanovo, 6 km from the Oka. From the town of Kasimov to the village of Popovka along the left bank, there is an elevated limestone slope. Up to the village of Popovka, it is treeless. On the right bank, there are flood meadows. Near the village of Maryino, they end. Before the confluence of the river Tashenka (11 km from Kasimov), there is the Zeleninsky quarry. Its anthropogenic slopes have an angle of 60-80°. The Upper Carboniferous limestone is mined here. Limestones of the Podolsky and Myachkovsky Carboniferous horizons of the Moscovian Age surround the Tashenka River.

Appendix III: Issues of the Oka Vegetation Distribution

On the left bank of the Oka, from the village of Popovka, a forest begins, at first broad-leaved, then a pine forest on a sandy terrace, which spreads to the village of Balushevo-Pochinki and further. The sandy terrace is located 2-3 km from the Oka. The area between the sandy terrace and the Oka watercourse is occupied by flood meadows (Figure 22).

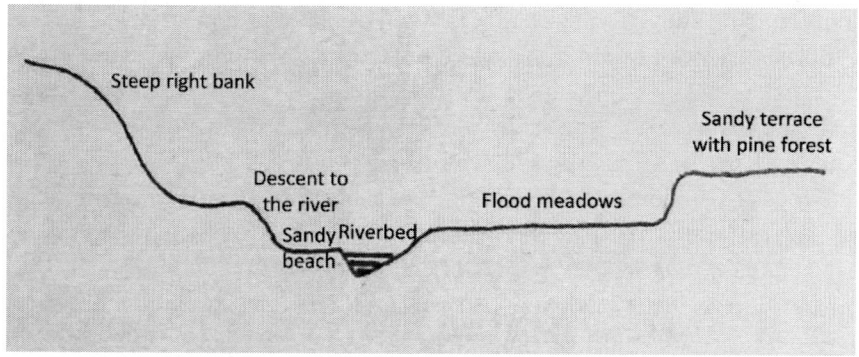

Figure 22. Schematic drawing of the Oka valley from the author's paper of 1974.

From the confluence of Tashenka and Oka to the village of Pervo, a birch forest grows along the slope of the Oka. Near Tashenka, it changes to pine and birch forest, and near the village of Pervo, it ends. Below the village of Maltsevo, the bank is covered with broad-leaved forest.

Next, we place the drawing of the landscape along the line of the village of Pervo – the village of Karamyshevo.

Species of southern flora are found along the Oka River on pine and alluvial sands, limestones (a distinction should be made between rocky slopes and slopes where the rock lies at some depth), and clays (alluvial soils). In places where flood meadows end and the sand terrace begins, there are also many species of steppe flora and shrub flora of the chernozem belt, e.g., ladybells [*Adenophora liliifolia* (L.) Bess], geranium (*Geranium sanguineum* L.), and skullcap (*Scutellaria hastifolia* L.).

On pine sands, the projective plant cover is ~ 30%, while in open places, it can be higher. Here, there are shrubs of the chernozem strip: brooms [*Cytisus Zingeri* (Nenuk.), V. Krecz., *C. rossicum* Fisch.], dyer's greenweed (*Genista tinctoria* L.). They make up a coverage of 5-10%. Turf cereals predominate: fescue [*Festuca sulcata* (Hack.) Nym.], blue hair grass [*Koeleria glauca* (Schrad.) DC]. In some places, they are found quite abundantly. Other specimens include evening primrose (*Oenothera biennis* L.), sand jurinea [*Jurinea cyanoides* (L.) Reichb.], valerian (*Valeriana wolgensis* Kasakew.).

The plants grown on the pine forest terrace represent the steppe flora; however, this cannot be fully explained by the relic hypothesis. The spread of *Oenothera biennis* L. already casts doubt on this hypothesis. As we know, this species came from North America. It has spread widely in the south of the European part of Russia, and to the north, it grows mostly along large rivers (mainly along the Oka). Oenothera is most often found on pine forest sands; it is not found on alluvial sands, and it is very rare on limestone slopes. The relic hypothesis cannot explain this since Oenothera was absent in Europe in the preglacial period.

Many plants of the line sandy terrace also prefer dry sandy places in the south (e.g., fescue often grows in semi-deserts).

Weeds often dominate the alluvial sands along the Oka. For example, on the banks of the Oka River in Kasimov, the following plants predominate: cocklebur (*Xantium strumarium* L.) predominates (up to 60-70% of the cover in some places), the species of goosefoot (*Chenopodiaceae*) family, amaranth (*Amaranthus retroflexus* L.). These species also predominate on alluvial sands near the village of Pervo. It should be pointed out that weeds appear only in places that are somehow changed by man. Near the village of Popovka the sands are almost completely free of impurities. Here one can find crabgrass [*Digitaria ischemum* (Schreb.) Muhl] (approx. coverage 30-35%), camel grass (*Corispermum Marschallii* Stev), and couch grass [*Elitrygia repens* (L.) Nevski]. In some places there are thickets of white marsh grass [*Petasites spurius* (Retz.) Reichb.].

In the town of Kasimov, the sands are littered with clay particles and other impurities. Near the village of Pervo, a plot of land was plowed near the sands. This field was abandoned, after which weeds began to grow around it: amaranth or pigweed, cocklebur, *Pulicaria prostrata* (Gilib.) Aschers. Near the village of Maryino, the sands are home to the Russian thistle (*Salsola pestifer* Nels.), which became an invasive plant in North America.

Taliev considered weeds as alien plants (Taliev, 1896). They inhabit alluvial sands and are found on limestone. Even on the anthropogenic slopes of the quarry, where there is almost no soil, weeds predominate. Here, grow such plants as sweet clover [*Melilotus albus* Desr., M. *officinalis* (L.) Lam.], wild carrot (*Daucus carota* L.), common groundsel (*Senecio vulgaris* L.), wild parsnip (*Pastinaca sylvestris* Garsault), and viper's bugloss (*Echium vulgare* L.). In some places, tansy (*Tanacetum vulgare* L.), prickly lettuce (*Lactuca serriola* Torner), yellow toadflax (*Linaria vulgaris* Mill.), white goosefoot (*Chenopodium album* L.), and wormseed wallflower (*Erysimum cheiranthoides* L.) can be found. The alien *Erysimum canescens* Roth was also found; it was approximately 1 m tall and brought many seeds. *Chaenorhinum viscidum* (Moench) Simk. is an alien plant that grows in relative abundance on the quarry slope. Tatar lettuce (*Lactuca tatarica* (L.) C.A. Mey) is also found on the slopes of the Oka. Weeds predominate on the rocky slopes of the Oka, but there are also meadow species (lotus, alfalfa, timothy grass, white bentgrass), which grow on the places where the soil has started to form.

On rocky areas of elevated slopes, there are also hemp-agrimony (*Eupatorium cannabinum* L.), tall wormseed wallflower (*Erysimum strictum* Gaertn, Mey et Scherb), and thyme-leaved dragonhead (*Dracocephalum thymiflorum* L.). These plants are typical of the Chernozem zone. Elevated limestone slopes may not be rocky. The parent rock may lie at some depth, but it still has a great influence on the soil composition. Such soil will have an alkaline pH, similar to the soils of the Chernozem zone.

Nonchernozemic soils are acidic. Many species cannot grow on them, and cultivated plants do not feel well; therefore, in a number of places, soil liming is used, i.e., the addition of calcium carbonate, after which the pH of the soil shifts towards alkalinity.

Species of the Chernozem zone have adapted to live on soil that gives an alkaline reaction and has a significant humus layer. Chernozem is an extremely fertile soil, which cannot be said about rocky outcrops. However, slopes in which limestone lies at a certain depth (3-45 cm) and have a substantial humus layer are more suitable for the growth of southern plants. Therefore, on such a slope along the Tashenka River near the village of Davydovo, more than 30 species of southern flora grow. The projective cover here is more than 90%, and the height of the upper tier is up to 80 cm. This slope of the Tashenka River is located 4 km from the Oka. Its slope is on average 10-25° but has also steeper areas. It has an exposition of the southwest-west, which favors the growth of steppe vegetation, opening the slope to the southern and western winds. There are also rocky outcrops on the slope, but the coverage here is small. The steppe plants that come here (for example, *Salvia pratensis*) have reduced vitality coefficients compared to meadows and glades, where limestone lies at a certain depth. On the northern half, the slope is overgrown with spruce and oak. Here, under the trees, the cover is 60-90%, dominated by *Agrostis Syreitschikovii*, *Carex montana*, *Calamagrostis arundinacea*, and *Melica picta*. Snowdrop anemone with large flowers (*Anemone sylvestris* L. f. *grandiflora* MB), feverfew [*Pyrethrum corymbosum* (L.) Willd.], and catnip (*Nepeta pannonica* L.) grow there.

On the southern part of the slope, spruce and birch grow (at the top, there are hazel and spruce). Southern forest plants are found in rare places under birch and pines: false-bromes (*Brachypodium pinnatum*, *B. sylvaticum*), hairy St. John's wort (*Hypericum hirsutum* L.). In the glades there are *Anemone sylvestris* L., *Inula hirta* L., *Potentilla recta* L., *Geranium sanguineum* L. and others. Sage (*Salvia pratensis* L.) dominates the meadow in the central part of the slope. There are many other southern plants. Of the cereals, timothy grass and cat grass predominate. The coverage in the meadows and glades is more than 90%.

Torilis japonica (Houtt.) DC and *Campanula bononiensis* L. are found in the thickets of hazel and spruce. They are widespread on the banks of the Oka, although they are still more common in the south. In the fields and meadows of the slope, many common meadow cereals and legumes grow, such as alpine clover, meadow clover, bird's-eye speedwell, yellow alfalfa, and bird's-foot trefoil.

The bare slopes of the Oka are not rich in vegetation. Here, cattle are chased everywhere, and the grass is trampled; the height of the grass is 10-20 cm, rarely more. On the treeless slopes, there are many plants of dry meadows: *Rumex acetosella* L., *Androsace septentrionalis* L., *Pilosella officinarum* F.W.Schultz & Sch.Bip., and *Ajuga genevensis* L. Plants of the southern regions are found only occasionally [*Nonea pulla* (L.) DC., *Elytrigia intermedia* (Host.) Nevski, *Dracocephalum thymiflorum* L.]. These are mainly the species of dry places of the black soil (chernozem) zone.

In the deciduous forests along the Oka, the Dutchman's pipe, *Aristolochia clematitis* L., is often found. Here, you can also see *Alexitoxicum officinale* St.-Lager. The limestone in these places lies deep, so there are no steppe plants here except *Aristolochia* and *Alexitoxicum*.

Clay soils predominate in the chernozem zone. In the non-chernozem zone, clay soils are found mainly in the floodplains of rivers and are of alluvial origin. Therefore, the soils of the flood meadows of the Oka are similar to those of chernozem in terms of mechanical composition. Both meadow and chernozem soils are very fertile, although the pH of meadow soils is lower than that of chernozem soils (Kosyakin, 1973). However, there is a similarity between these soils, so many species of grassy steppes grow in flood meadows: asparagus (*Asparagus officinalis* L.), burnet (*Sanguisorba officinalis* L.), bulbous bluegrass (*Poa bulbosa* L. var. *vivipara* Koel.), and noble yarrow (*Achillea nobilis* L.).

Main Findings

1. Humans play a major role in the spread of plants. They have stripped the banks of rivers, cut down forests, and drained meadows. Unpretentious southern plants, mostly weeds, populate these places.
2. The relic hypothesis of the origin of the Oka vegetation has many errors. Even in areas not touched by humans (meadows, wooded slopes), vegetation was brought along the Oka mostly naturally and less often due to humans. The relict hypothesis cannot explain the vegetation of the Oka.
3. Weed vegetation is mostly alien. Many species have been brought in from Canada and Central Asia.
4. Now weeds are conquering flood meadows, which are being improperly developed everywhere.
5. The acidity and composition of the Oka soils determine the growth of steppe flora species on them.

References

Kaufman, N. N. (1866). *Moscow Flora, or Description of Higher Plants and Botanical-Geographical Essay of Moscow Province.* Publishing house of A.I. Glazunov, Moscow.

Kosyakin, A. S. (1973). *Oka Meadows.* Moskovsky Rabochiy, Moscow.

Litvinov, D. I. (1899). On the Oka flora. *Materials for the Knowledge of the Fauna and Flora of the Russian Empire* 3, 16–17.

Litvinov, D. I. (1902). On the relict nature of the flora of the stony slopes of European Russia. *Proceedings of the Botanical Museum of the Russian Academy of Sciences 1,* 76–109.

Taliev, V. I. (1895). Chalk Forests of Donets and Volga Basins. *Proceedings of the Society of Naturalists at Kharkiv University* 29, 225–282.

Taliev, V. I. (1896). An outline of weed biology. *Natural Science and Geography* 8, 812–821.

Index

A

agriculture, 47, 65, 66, 69, 70, 73, 97
algae, 7, 8, 19, 31, 37, 38, 39, 42, 49, 93, 95
anthropocene, 56, 69, 75, 89
anthropogenic factor, 62, 63, 64, 66
archaean eon, 5, 6, 7, 15

B

Baishevsky settlement, 70, 73, 77
Bashkirian age, 5, 16, 18
brachiopods, 13, 14, 20, 29, 30, 56

C

calamite, 26
Cambrian period, 7, 9, 11
Carboniferous period, 5, 10, 13, 14, 15, 16, 18, 20, 21, 23, 25, 29, 30, 31, 35, 45, 50
Cenozoic era, 32, 36, 46, 47, 49
civilization, xi, 1, 4, 52, 55, 56, 75, 76, 85
CO_2, 23, 24, 33, 34, 36, 37, 38, 39, 40, 41, 42, 43, 44, 45, 46, 47, 48, 49, 52, 75, 87, 88, 89, 90, 91, 93, 94, 95, 96
corals, 8, 13, 14, 32

D

Danshin, Boris, 16, 17
Dashkov, Fyodor, 54, 55, 63, 71, 72, 85, 97, 103
Devonian period, 10, 11, 35, 48
Dvigubsky, Ivan, 59

E

East European platform, 5, 9, 10, 15, 20, 85, 90
evolution, xi, 4, 9, 26, 29, 34, 35, 37, 38, 40, 45, 46, 49, 51, 64, 66, 85, 87, 88, 90, 91, 92, 93, 94, 95

F

fern, 25, 27
floodplain, 1, 54, 55, 60, 70, 76
foraminifera, 13, 14, 19, 31, 36, 93

G

Gagin, Ivan, 70, 72, 79, 81, 82
glacial, 13, 16, 18, 19, 24, 36, 39, 43, 44, 49, 51, 52, 53, 54, 61, 75, 89, 94, 110
glacial-interglacial oscillations, 19, 43, 49, 51
greenhouse burst, 47, 48

H

Holocene, 52, 53, 55, 56, 57, 65, 69, 70, 76, 90
Holocene calendar, 55, 69

I

interglacial, 16, 18, 24, 47, 51, 53, 94
Iron Age, 63, 71

J

Joggins Fossil Cliffs, 14, 26, 27, 90

K

Kasimov, 1, 3, 5, 10, 13, 14, 15, 16, 18, 19, 20, 29, 30, 32, 51, 54, 55, 56, 63, 69, 70, 71, 73, 76, 77, 78, 79, 80, 81, 82, 83, 85, 89, 97, 98, 99, 103, 106, 108, 109, 110, 112
Kasimovian Age, 5, 13, 15, 16, 18, 19, 24, 26, 27, 29, 56
Kaufman, Nikolai, 59, 60, 61, 62, 110, 115

L

land plants, 13, 33, 34, 36, 38, 39, 41, 42, 43, 44, 45, 48, 49, 88, 89, 91, 92
limestone, 1, 4, 5, 13, 15, 18, 19, 20, 30, 31, 36, 49, 51, 56, 59, 61, 63, 64, 65, 70, 71, 75, 76, 77, 78, 79, 100, 101, 110, 112, 113, 114
Litvinov, Dmitry, 61, 62, 110, 115

M

Mansurov, Alexei, 71, 73, 83, 103
Mayevsky, Pyotr, 59, 61
Mesozoic era, 15, 30, 36, 42, 48, 49
methane, 6, 7, 9, 33, 37, 47, 52, 54, 75
Moscovian Age, 16, 18, 19, 20, 24, 26, 56, 110

N

neolithic, 1, 2, 54, 65, 69, 70, 71, 72, 73, 76, 91, 98
Neolithic settlement, 2, 69, 71, 72, 98
Nikitin, Sergei, 16, 17

O

O_2, 24, 33, 34, 35, 36, 37, 38, 39, 40, 41, 42, 43, 44, 45, 46, 47, 48, 49, 87, 88, 91, 93

Oka flora, 57, 59, 60, 61, 62, 63, 65, 67, 78, 91, 109, 110, 115
Oka River, xi, 1, 2, 27, 57, 59, 60, 63, 78, 79, 80, 81, 111, 112
Ordovician period, 30, 35
Oxbow Lake, 56, 57

P

paleolithic, 54
Paleozoic era, 7, 15, 18
Permian period, 15, 18, 23, 24, 30, 45
Pervo (village), 1, 2, 3, 19, 56, 67, 70, 72, 76, 77, 78, 85, 97, 98, 99, 101, 103, 105, 106, 107, 111, 112
phanerozoic eon, 7, 8, 34, 36, 40, 44, 46
photorespiration, 36, 37, 40, 41, 42, 43, 45, 46, 48, 49, 94
photosynthesis, 6, 7, 13, 15, 33, 34, 37, 38, 39, 41, 42, 43, 49, 87, 89, 90, 91, 92, 93, 94, 95
phytospreading, 64
Platonov, Andrei, 4
Pleistocene, 15, 44, 45, 48, 49, 55
Proterozoic eon, 5, 7, 9, 36, 39, 91

Q

quarry, 55, 70, 71, 77, 78, 110, 112
Quaternary period, 16, 19, 20, 51, 65

R

Relict flora, 62
respiration, 40, 42, 43, 87, 88, 89

S

Sea Lilies, 13, 14, 19, 30
Silurian period, 35
Smirnov, Pavel, 60

T

Taliev, Valery, 61, 62, 63, 64, 65, 66, 67, 78, 90, 94, 110, 112, 115

Tashenka River, 2, 54, 56, 63, 67, 71, 77, 97, 110, 113
Tatar culture, 79
Trapa natans, 56, 57

W

weathering, 10, 15, 18, 34, 39, 41, 42, 43, 48, 49, 88, 92, 93

wheatgrass, 73, 87, 92, 95

Z

Zagoskin, Mikhail, 2, 4, 70, 95
Zinger, Vasily, 62